SpringerBriefs in Applied Sciences and Technology

Forensic and Medical Bioinformatics

Series editors

Amit Kumar, Hyderabad, India
Allam Appa Rao, Hyderabad, India

More information about this series at http://www.springer.com/series/11910

Mohammad Mostakhdemin · Iraj Sadegh Amiri
Ardiyansyah Syahrom

Multi-axial Fatigue of Trabecular Bone with Respect to Normal Walking

 Springer

Mohammad Mostakhdemin
University Technology Malaysia
Skudai
Johor
Malaysia

Ardiyansyah Syahrom
University Technology Malaysia
Johor Bahru
Malaysia

Iraj Sadegh Amiri
Photonics Research Centre (PRC)
University of Malaya (UM)
Kuala Lumpur
Malaysia

ISSN 2191-530X   ISSN 2191-5318   (electronic)
SpringerBriefs in Applied Sciences and Technology
ISSN 2196-8845   ISSN 2196-8853   (electronic)
Forensic and Medical Bioinformatics
ISBN 978-981-287-620-1   ISBN 978-981-287-621-8   (eBook)
DOI 10.1007/978-981-287-621-8

Library of Congress Control Number: 2015942489

Springer Singapore Heidelberg New York Dordrecht London

Springer Science+Business Media Singapore Pte Ltd. is part of Springer Science+Business Media
(www.springer.com)

Abstract

Weakened trabeculae are one of the most important clinical demands which need to be analysed to prevent fracture of hip joint. This analysis supports some goals clinically and mechanically, such as treatment of osteoporotic bone based on drug delivery and tracking the fatigue behaviour of bone or considering interaction of implants with trabecular bone mechanically. To find bone material behaviour that is faced with different types of loads make a useful contribution to construct artificial bone. In this study a sample taken from bovine trabecular bone selected and osteoporosis disease simulated was subjected to axial, torsional and multi-axial load and correlation of load amplitude and their type with number of cycles to failure was found. Struts found that very weak against torsional load and damage initiated at the most arch shape of structure. Plastic localization considerably increases in torsional and multi-axial load with respect to axial load. Volumetric elastic strain and effective plastic strain were computed numerically in each type of load as static analysis then results in transfer to dynamic analysis part for fatigue simulation. S–N curve extracted for all three types of load show that axial load had a considerable fatigue life which reached 1226 cycles in 40 % of the total load due to normal walking. In contrast, this value for torsional load and multi-axial load reached to 42 and 38 cycles respectively. Fatigue life of bone decreases drastically in osteoporotic bone when torsional load has been imposed during physiological activities.

Contents

Chapter 1
Introduction of Bone Study

Abstract Trabecular bone plays an important role in skeleton structures. Some analysis such as numerical, analytical or experimental analysis applies, to predict fatigue life of trabecular bone. In this study, numerical analysis is performed to predict fatigue life of trabecular bone subjected to axial compression mixed with torsional load due to gait loading. Objective of study is established to predict fatigue life of bone structure. Considerably, trabecular bone makes useful contribution all over the body because of load tolerating duty. Effect of multi-axial and discrete load due to normal walking is highly demanded clinically. The scope of this study is to compare the fatigue behaviour of trabecular bone faced with multi-axial and discrete load. Analysing fatigue life of trabecular bone based on average bodyweight in this decade; make a useful contribution for human being and their health, especially for those who suffered from osteoporosis disease.

1.1 Introduction

Bone is known as the most important mechanism of body that tolerate internal and external load due to weight of body or physiological activities. Since there are movements in body, some parts such as bones, joints and muscles are involved in this investigation. Among such these parts, trabecular bone plays an important role in skeleton structures. Duty of bone is not limited to only carry the most percentage of loads; even it is a secure atmosphere for blood vessels and marrow in its inner sites. Inner site of bone, which like spongy structure called trabecular bone, tolerate most percent of bodyweight. Some analysis such as numerical, analytical or experimental analysis applies, to predict fatigue life of trabecular bone. In this study, numerical analysis is performed to predict fatigue life of trabecular bone subjected to axial compression mixed with torsional load due to gait loading.

© The Author(s) 2016
M. Mostakhdemin et al., *Multi-axial Fatigue of Trabecular Bone with Respect to Normal Walking*, Forensic and Medical Bioinformatics,
DOI 10.1007/978-981-287-621-8_1

When load is imposed on hip, because of its especial shape, load is transmitted to trabecular bone and different anatomical sites faced with different load angle. In this study, load imposed on femoral head of bovine bone which its direction is parallel with neck of femoral bovine bone was selected. Stress analysis first is performed, and then fatigue analysis and number of cycles to failure data for trabecular bone is extracted. This chapter covers the problem background, problem statement, objectives and scope of study.

1.2 Study of Fatigue Behaviour of Trabecular Bone Subjected to Multi-axial Load

In this book objective of study is established to predict fatigue life of bone structure, to consider reaction of trabecular architecture that was faced with compression and torsional load to extract fatigue S-N curve respect to normal walking. Trabecular bone subjected to axial compression and torsional loading investigated and the relation of results with bone morphology is pointed. Finally, Strain versus number of cycles to failure for model will be analysed, range of fatigue life can present correlation of fatigue life of bone and morphological indices which will be compared with previous study.

1.3 Effects of Various Loads on Trabecular Bone

These days, artificial architecture of parts of body has a priority in academic research with which, find some methods for analysing bone structure and being closed to real structure in the sense of mechanical properties. These methods are applied to investigate mechanical properties of bone to analysis of some physics of bone such as fatigue life of considered structure and its behaviour respect to physiological activities, and then compare fatigue data extracted by various type of loads to find out effect of each load on it.

Considerably, trabecular bone makes useful contribution all over the body because of load tolerating duty. However, many percent of stress distribution are tolerated by trabecular bone. This study is point out to identify, how the stresses due to physiological activities distribute over the bone and to what extend these stresses influence on fatigue life. Fatigue life prediction based on combination of axial compression and torsional load of trabecular bone will be covered. Effect of multi-axial and discrete load due to normal walking is highly demanded clinically.

1.4 Research Scope on Fatigue Analysis of Trabecular Bone

First scope of this study is to reconstruct the trabecular bone structure in effective quality to prepare it for mechanical analysis. Since there is high-tech system such as microcomputer tomography (micro CT-Scan) scanner, which assists to construct the complex structures by using especial software called Materialize mimic software. Mimic is able to have link with those images taken from Micro-CT scan and construct any 3D-complex structure such as trabecular bone which causes trustworthy results in the pre-processing step in FE software that strongly influence in the final results.

The second scope of this study is, fatigue analysis which would be done by numerical, analytical and experimental approach. Researchers made their effort to use some methods to get their results. Since experimental approach has limitation of imposing load compare with the actual condition and limitations of machine and mechanism movements and measuring instrumentations when multi-axial load must be considered, thus, finite element method (FEM) is reliable method that can be applied to simulate fatigue life of trabecular bone.

Third scope of this study is to compare the fatigue behaviour of trabecular bone faced with multi-axial and discrete load. Analysis of multi-axial load which imposed on trabecular bone in body respect to physiological activities clear the effect of various types of loads on bone fatigue life and this is the way to get idea for construct idealize structure and use it in the body as artificial structure or some treatments based on drug delivery which is aimed to know weaken parts of bone faced with osteoporosis diseases.

1.5 Contribution of Study in Fatigue Life of Bone

The first significance of this study is considering the mechanical properties in common physiological activities. Gait loading is known as one of the common activities that everyone is involved with it all the days; In addition, due to the fast life, common diseases among people are common such as obesity. Analysing fatigue life of trabecular bone based on average bodyweight in this decade; make a useful contribution for human being and their health, especially for those who suffered from osteoporosis disease.

The second significance of this study is to know to what extend various types of loading due to various daily activities strongly influence on the bone structures and which of them cause do damage more. Among various loading, cyclic loading include of axial, torsional and multi-axial load play an important role to damage bone.

Fatigue life prediction of trabecular bone and using FEM cause estimate the crack initiation and crack growth and its location based on strain accumulation analysis; there are current problems which should investigate according to the common physical activities. Osteoporotic fracture also is one of the famous diseases occurring due to the excessive loads on the bone. Consequently, prediction of fatigue life of bone samples can make clear the way of implants and prosthesis construction.

Chapter 2
Literature Review Fatigue Analysis in Trabecular Bone

Abstract Normal walking is the most action impose on the skeleton structure. The microarchitecture of trabecular bone plays an important role respect to mechanical properties. Analyse the fatigue behaviour of the trabecular bone respect to physiological activity (Normal Walking), subjected to combination of axial compression and torsional (multi-axial) load counted as the main aim of this study. The osteoclast is responsible for modelling and remodelling of bone and is defined as a large multinucleate bone cell that absorbs bone tissue during growth and healing. Irregularities and disorders in trabecular bone cause to reduction of bone mass and its architecture. The standard method applied to extract bone structures properties is 2D section of bone biopsies. Tetrahedrons technique is applied to calculate bone volume (BV), total volume (TV) is the volume of whole bone structures. Trabecular bone has a significant portion in respect of resisting compression and shear. Data extracted from experimental test is depends on many parameters such as geometry of bone and measurement of strain. Trabecular architecture has a specific properties respect to tension loading. Almost all load due to physiological activates are counted as cyclic loading. Lifetimes were found to be highly dependent on the axis of loading and are drastically reduced for off-axis loading.

2.1 Introduction

Since skeleton play an important role on supports weight of body, analysis of mechanical properties is highly recommended to identify weaken parts due to physiological activities, however, among daily activities normal walking is the most action impose on the skeleton structure. Human bone is categorized in two various types, the first is called Cortical bone which has high porosity and density, the second one that is inner part of bone structure is called trabecular bone or Cancellous as commercial name.

© The Author(s) 2016
M. Mostakhdemin et al., *Multi-axial Fatigue of Trabecular Bone with Respect to Normal Walking*, Forensic and Medical Bioinformatics,
DOI 10.1007/978-981-287-621-8_2

Since daily activities, bone structures face with various types of loading with which trabecular bone tolerates approximately 70 % of total load [1]. The micro-architecture of trabecular bone plays an important role respect to mechanical properties. To analyse samples constructed by micro-CT scan images, finite element model was applied for failure mechanism and local strain field.

Damage accumulation is a major factor of weakens vertebrae due to cyclic loading and is also cause of failure in implants [2]. Among many researches that covers fatigue analysis, fatigue parameters such as fatigue strength in cortical parts is extensively reported [3–6] however, fewer data are reported for trabecular part [7]. Analyse the fatigue behaviour of the trabecular bone respect to physiological activity (Normal Walking), subjected to combination of axial compression and torsional (multi-axial) load counted as the main aim of this study.

2.2 Bone Rehabilitation Mechanism

Since skeleton has duties as framework of body to tolerate load and stresses due to daily activities, remodel and reshape itself by some cells in a cyclic period of time. Bone consists of organic matrix, inorganic minerals, cells and water that resist mechanical forces.

2.2.1 Bone Modelling and Remodelling

Bone remodelling (bone metabolism) is a process which is done by cell called the osteoclast that diminish mature bone and another cell called osteoblast which has a duty of rehabilitation of those spaces that faced with osteoclast and this process is called ossification. This process is completely varied among different ages, such a process is done 100 % for first year of life, however, for adult it will decreases to 10 % totally. Remodelling process and its effect on bone can be counted as a mechanical function, and this condition of bone structures can be analysed based on human activities or any fracture, thus, micro damage occurred in bone.

2.2.1.1 Osteoblast Cells in Bone

The osteoblast supports bone in the sense of producing more cells matrix, termed osteoid. In adulthood, osteoid exists on the surface of trabecular bone, however, on the inner lining of cortical bone.

2.2.1.2 Osteoclast Cells in Bone

The osteoclast is responsible for modelling and remodelling of bone and is defined as a large multinucleate bone cell that absorbs bone tissue during growth and healing.

2.3 Bone Architecture and Materials

Bone is divided into two types, the one is immature and the second is mature. Mature bone is formed by immature one, and following its remodelling. Mature bone is divided into two types: Trabecular bone or cancellous bone and compact or cortical bone. Trabecular bone has a space with marrow cavities, which is a space for vessels and mineral nutrition pass; on the other hand, compact bone is more rigid respect to cancellous bone and include of the marrow channel. Cortical bone has enough strength for withstand of weight and forces imposed by physiological activities, however, trabecular bone is more flexible and reservoir for calcium homeostasis. By become older and older, portion of trabecular bone diminish. Trabecular form a network of rod and plate-like elements that suitable for blood vessels pass and more lighten than cortical bone with marrow cavities. Detail is illustrated in Fig. 2.1 [8].

Fig. 2.1 Trabecular bone

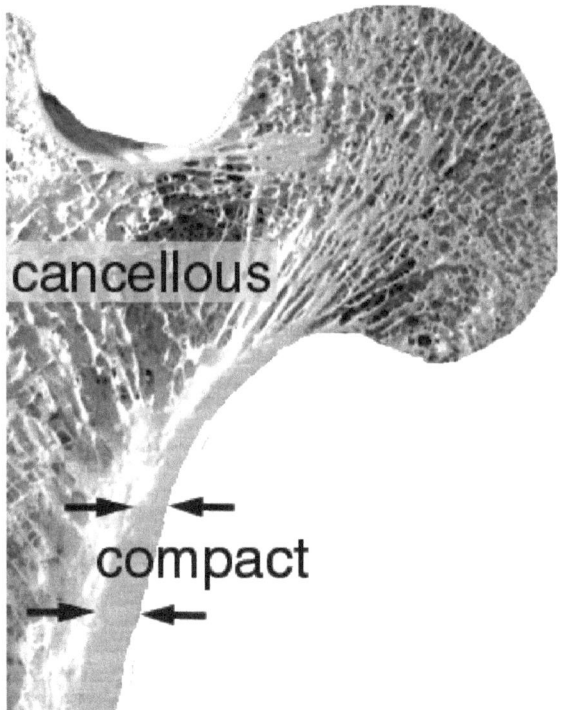

The trabecular bone is made up of trabecular plates and rods (struts and plates). Whilst cortical bone can be regarded as homogeneous and isotropic, in comparison of trabecular bone that is anisotropic and inhomogeneous [9] has variations in its structures depends on different anatomical sites [10].

2.4 Morphological Indices of Bone

Irregularities and disorders in trabecular bone cause to reduction of bone mass and its architecture, 94 % is counted as trabecular strength if bone density and architectural are measured, furthermore, use of bone density, 64 % is counted as its portion alone. The standard method applied to extract bone structures properties is 2D section of bone biopsies. In addition, three dimensional morphology indices will be extracted form 2D images that this technique is called stereological methods. Some morphological indices such as bone volume (BV/TV) and surface density (BS/BV) are extracted from samples, and other crucial data such as trabecular thickness (Tb. TH), trabecular separation (Tb. SP), and trabecular number (Tb. N) are extracted indirectly if structure is assumed fixed part, but trabecular bone is faced with changes in its shape and architecture within different times, so this assumption will lead to have error on extracted indices [11].

Marching Cubes method (MCM) is used to calculate bone surface area (BS) in this methods surface being triangulated of mineralized bone phase. Tetrahedrons technique is applied to calculate bone volume (BV), total volume (TV) is the volume of whole bone structures. BV/TV and BS/TS are used to compare samples with different architecture.

2.5 Bone Mechanical Properties

Since many reasons, studies of mechanical properties of bone make it useful for orthopaedic science. First, knowledge of this properties will clearance what behave is expected from bone in life and to what extend bone has abilities for absorbing energy and so on. Bone is known as composite materials, include of minerals, waters and cells.

According to the different sites, ages and diet, this composition is differed. In this composition, 90 % of bone matrix is collagen. Collagen has a low modulus, poor compressive strength and excellent strength in tensile. Hydroxyapatite is mineral phase of bone that is stiff and with good tensile strength regards of mechanical aspects. Combinations of these two materials make bone as anisotropy material that is strong in compression but weak in shear, however, in respect of tension is intermediate compare to another [12].

Trabecular bone has a significant portion in respect of resisting compression and shear, also trabecular bone get 25 % as dense, 500 % as ductile and 10 % as stiff as cortical bone. Based on different position and ages, trabecular has different mechanical properties, it is open-celled porous foam and combination of rod and plate–like that depends on architecture and orientation of those rod and plate-like, mechanical properties varies functionality [12].

2.5.1   Static Properties of Bone

2.5.1.1   Compression Properties of Trabecular and Cortical Bone

Many studies report mechanical compressive characterization of the human trabecular bone. Compressive properties and their relations with the trabecular bone density and morphological parameters are well known. Most of these studies aim to predict the trabecular bone strength in normal in vivo loading; however, trabecular bone transmits essentially compressive and tensile loads and due to this transition trabecular is faced with multi axial stresses [13]. Since trabecular bone mechanical testing is not as convenient as cortical bone, and based on some experimental research, mechanical properties of trabecular bone is close to cortical bone, for example, Young's modulus for trabecular bone is 20 % lower than cortical bone [14]. Data extracted from experimental test is depends on many parameters such as geometry of bone and measurement of strain [15, 16]. Failure properties of trabecular bone in compression test cannot be exactly explained with established material properties of cortical one [17].

Some research carried out on uniaxial and confined compression proof that hydrostatic yield stress with uniaxial yield stress for trabecular bone is equal, further pressure dependent plasticity is more accurate for simulation [18]. In addition, elastic modulus and yield strain of trabecular bone is reported lower than cortical bone and this cause to cumulative effect between cortical and trabecular tissue with which tissue strength for cortical bone is 25 % greater than trabecular tissue [19]. Compression loading to reach yield strain in on-axis loading and off-axis has been performed, yield strain increased in off-axis loading and reduction in strength that is related to the off-axis loading is greater than modulus [9].

According to the compression and torsion loading, powers of density in compression is larger than in torsion, however strain rate has larger value in torsion than compression. In addition, for different trabecular shear properties, effect of density is weaker compared with effect of strain rate which is stronger [20]. In this study they performed that in compression loading the changes in bone volume cause compress marrow and this phenomenon is counted as reason of bone stiffening increase within compression loading, the power of density has significant contribution in compression than those in torsion. On the other hand, ultimate strength, yield strength and bone stiffness have a close relation in compression loading for trabecular bone, finite element analysis is applied to measure the bone strength by estimation stiffness [21].

2.5.1.2 Torsion Properties of Trabecular Bone

Torsion and shear properties are significantly correlated respectively with apparent densities of torsion and shear specimens [22]. Some parameters such as damage shear modulus, shear yield stress and ultimate shear stress has are related to induce damage to analysis of bone strength in especial disease such as osteoporosis. Changes in ultimate shear strength and toughness are proportional to decrease shear modulus, these two factors are more susceptible in diminishing volume fraction [23].

First of all measuring elastic properties of high degree porosity structure such as trabecular bone is difficult with traditional methods, the best technique of measuring Young's modulus and shear modulus is using ultrasonic technique [24]. Some researches carried out on the torsional properties, effect of trabecular bone in two various group such as bone with marrow and without were investigated, [20] reported applied these groups in low strain rate (0.002 s^{-1}) and high strain rate (0.05 s^{-1}), power relation to determine shear strength and shear modulus has been applied as shown in Eqs. 2.1 and 2.2 [25].

$$\sigma_u = 40.2 \times \rho^{1.65} \times \dot{\varepsilon}^{0.073} \tag{2.1}$$

$$E = 2232 \times \rho^{1.56} \times \dot{\varepsilon}^{0.047} \tag{2.2}$$

In this report, shear strength was proportional to the density raised to the 1.02 power and strain rate was raised to the 0.13 powers. In addition the shear modulus is proportional to apparent density raised to 1.08 and strain rate raised to 0.07 powers. However, [23] performed decreasing in shear modulus cause to changes toughness and shear strength.

2.5.1.3 Calculation of Shear Stress and Shear Strain for Trabecular Bone

After preparation of bone sample for torsional test in the sense of using water jet to remove all marrow and filled samples up by PMMA for its fixation part and put it in the brass shaft to hold sample, torque (N m) versus deformation θ (°) curve will be obtained. The linear region of such a curve is torsional stiffness which is defined by $K = \frac{\Delta T}{\Delta \theta}$ [20]. Then shear modulus calculated as

$$G = \frac{K \times L}{J} \tag{2.3}$$

where T is the applied torque, L is the gage length and J is the polar moment of inertia. It is possible to consider trabecular bone as a continuum structure if more than five intra trabecular spaces exist within its dimensions [26, 27]. Our sample has more than 5 intra trabecular so it is possible to consider it as a continuum structure.

Shear stress is calculated by using torque-angle diagram and following equation reported [28]. Shear stress equation is as shown below:

$$\tau = \frac{1}{2\pi R^3}(\emptyset \frac{dT}{d\emptyset} + 3T) \tag{2.4}$$

where $\emptyset$ is the angular deformation per unit length ($\emptyset = \left(\frac{\theta}{L}\right)$) and R is the specimen radius. This equation is appropriate for approximately transversely isotropic trabecular bone samples.

To find maximum shear strength, consideration of the maximum point (peak) of torque-deformation diagram where $\frac{dT}{d\emptyset} = \frac{dT}{d\theta} = 0$ gives maximum shear stress as formulated below [20].

$$\tau_{max} = \frac{3T_{max}}{2\pi R^3} \tag{2.5}$$

Shear strain rate also is calculated as

$$\dot{\gamma} = \frac{R \times \dot{\theta}}{L} \tag{2.6}$$

which $\dot{\theta}$ is the deformation rate; with these two equations maximum shear stress and strain on the surface of bone would be calculated.

Correlation between strain rate increasing and significant increase in samples shear modulus and shear strength reported in [20]. They performed that presence of bone marrow did not effect on the shear modulus and shear strength. The power of density has significant contribution in compression than those in torsion; however, the power of strain rate is larger in torsion in respect to the compression.

2.5.1.4 Tensile Properties of Trabecular Bone

Trabecular architecture has a specific properties respect to tension loading. Apparent yield strain versus bone properties such as volume fraction is counted as function of each other in tension and compression loading, which yield strain remains constant within an anatomic site. In the axial loading, variation of yield strain in tissue level and apparent level has connectivity with each other, however, for bending loading is not [29]. Apparent yield strain variation within an anatomic site is too small and this result has been investigated [30].

Trabecular bone subjected to a bending load will face with high tensile stresses that would effect on strain localization within the bone. Tensile loading condition is assumed to be following Woff's law. When load direction is normal to the principal trabecular orientation, neither the compressively nor the tensile strain regions were clearly elongated, but they were generally within the transverse plane perpendicular to the vertical trabecular. Moreover, the regions were less than one half of the mean thickness of the trabecular struts, and the number and mean volume of the yielded regions both increased uniformly with increasing apparent strain. Together these findings are consistent with bending of the trabecular and the formation of plastic or damage hinge [31]

2.6 Fatigue Behaviour of Trabecular Bone

Almost all load due to physiological activates are counted as cyclic loading. These days, damage of trabecular tissue due to repeated loading is highly demand area in aspect of biological analysis. Cyclic loading cause damage and initiate crack, even though the load and stress amplitude is far below yield strength. Cyclic failures of bone due to accumulation of plastic strain is known clinically as overuse injuries or stress fractures. Cyclic loading and damage is reported to weaken vertebrae [32].

Lifetimes were found to be highly dependent on the axis of loading and are drastically reduced for off-axis loading. Also strains at failure showed to be a function of the deviation from the physiological axis which may reject the assumption of isotropic failure strains [1] The significant increase in life when material degradation is included may be understood by examining the various parameters (stress, modulus degradation and plastic strain) at a local level. Within the structure, material degradation was limited to highly localized regions around the areas of peak stresses. Due to the intrinsic weakness of bone in tension, material degradation was initiated in the regions of high tensile stresses.

Once a trabecular had completely fractured, stresses were redistributed to nearby trabecular, which experienced greater damage rates, and rapid failure of the sample then occurred. At a localized level, within the early load cycle's material degradation caused a reduction in the peak tissue stresses. During the fatigue life, significant modulus degradation and permanent strains were observed in these high stress regions. However, the volume of material undergoing material degradation only accounted for a small percentage of the total bone volume. On a global level, significant modulus degradation only occurred once a trabecular had fractured. Also, the accumulated permanent strain at failure was typically 10 % of the initial applied apparent strain. As the initial apparent strain level was increased, there was a trend of increasing the degree of modulus degradation and accumulated permanent strain just prior to failure [33].

Modulus reduction and specimen residual strain increased when maximum compressive strain increase. The post-test mechanical properties were most depends on maximum compressive strain and suggested that trabecular bone failure is largely strain based [34].

References

1. Dendorfer, S., Maier, H. J., & Hammer, J. (2009). Fatigue damage in cancellous bone: an experimental approach from continuum to micro scale. *Journal of the Mechanical Behavior of Biomedical Materials, 2*(1), 113–119.
2. Bauer, J. S., et al. (2007). Analysis of trabecular bone structure with multidetector spiral computed tomography in a simulated soft-tissue environment. *Calcified Tissue International, 80*(6), 366–373.
3. Carter, D. R., et al. (1981). Uniaxial fatigue of human cortical bone. The influence of tissue physical characteristics. *Journal of Biomechanics, 14*(7), 461–470.
4. George, W. T., & Vashishth, D. (2006). Susceptibility of aging human bone to mixed-mode fracture increases bone fragility. *Bone, 38*(1), 105–111.
5. O'Brien, F. J., Taylor, D., & Lee, T. C. (2003). Microcrack accumulation at different intervals during fatigue testing of compact bone. *Journal of Biomechanics, 36*(7), 973–980.
6. Yeni, Y. N., et al. (2009). Human cancellous bone from T12-L1 vertebrae has unique microstructural and trabecular shear stress properties. *Bone, 44*(1), 130–136.
7. Moore, T. L. A., O'Brien, F. J., & Gibson, L. J. (2004). Creep does not contribute to fatigue in bovine trabecular bone. *Journal of Biomechanical Engineering, 126*(3), 321–329.
8. Whitehouse, W., & Dyson, E. (1974). Scanning electron microscope studies of trabecular bone in the proximal end of the human femur. *Journal of Anatomy, 118*(Pt 3), 417.
9. Bevill, G., Farhamand, F., & Keaveny, T. M. (2009). Heterogeneity of yield strain in low-density versus high-density human trabecular bone. *Journal of Biomechanics, 42*(13), 2165–2170.
10. Kadir, M. R., Syahrom, A., & Ochsner, A. (2010). Finite element analysis of idealised unit cell cancellous structure based on morphological indices of cancellous bone. *Medical and Biological Engineering and Computing, 48*(5), 497–505.
11. Tor Hildebrand, A. L., 1 Ralph mü, L., 2 Jan D., & 3 Peter Rü E.1. (1999). Direct three-dimensional morphometric analysis of human cancellous bone: Microstructural data from spine, femur, iliac crest, and calcaneus. *Journal of Bone and Mineral Research, 14*.
12. Sudheer Reddy, M. D., & Soslowsky, L. J. (2009). *Biomechanics—Part I*. Berlin: Springer. doi:10.1007/978-1-59745-347-9_3.
13. Brown, T. D., & Ferguson A. B. Jr. (1980). *Mechanical Property Distributions in the Cancellous Bone of the Human Proximal Femur* (Vol. 51). London: Informa Healthcare.
14. Taylor, M. J. C. & Zioupos, P. (2002) *Finite element simulation of the fatigue behaviour of cancellous bone* (Vol. 37, p. 419). Springer link.
15. Linde, F., Hvid, I., & Madsen, F. (1992). The effect of specimen geometry on the mechanical behaviour of trabecular bone specimens. *Journal of Biomechanics, 25*(4), 359–368.
16. Keaveny, T. M., et al. (1993). Theoretical analysis of the experimental artifact in trabecular bone compressive modulus. *Journal of Biomechanics, 26*(4), 599–607.
17. Verhulp, E., et al. (2008). Indirect determination of trabecular bone effective tissue failure properties using micro-finite element simulations. *Journal of Biomechanics, 41*(7), 1479–1485.
18. Kelly, N., & McGarry, J. P. (2012). Experimental and numerical characterisation of the elasto-plastic properties of bovine trabecular bone and a trabecular bone analogue. *Journal of the Mechanical Behavior of Biomedical Materials, 9*, 184–197.
19. Bayraktar, H. H., et al. (2004). Comparison of the elastic and yield properties of human femoral trabecular and cortical bone tissue. *Journal of Biomechanics, 37*(1), 27–35.
20. Kasra, M., & Grynpas, M. D. (2007). On shear properties of trabecular bone under torsional loading: effects of bone marrow and strain rate. *Journal of Biomechanics, 40*(13), 2898–2903.
21. Fyhrie, D. P., & Vashishth, D. (2000). Bone stiffness predicts strength similarly for human vertebral cancellous bone in compression and for cortical bone in tension. *Bone, 26*(2), 169–173.

22. Follet, H., et al. (2005). Relationship between compressive properties of human os calcis cancellous bone and microarchitecture assessed from 2D and 3D synchrotron microtomography. *Bone, 36*(2), 340–351.

23. Garrison, J. G., Gargac, J. A., & Niebur, G. L. (2011). Shear strength and toughness of trabecular bone are more sensitive to density than damage. *Journal of Biomechanics, 44*(16), 2747–2754.

24. Ashman, R. B., Corin, J. D., & Turner, C. H. (1987). Elastic properties of cancellous bone: Measurement by an ultrasonic technique. *Journal of Biomechanics, 20*(10), 979–986.

25. Linde, F., et al. (1991). Mechanical properties of trabecular bone. Dependency on strain rate. *Journal of Biomechanics, 24*(9), 803–809.

26. Harrigan, T. P., et al. (1988). Limitations of the continuum assumption in cancellous bone. *Journal of Biomechanics, 21*(4), 269–275.

27. Kasra, M., & Grynpas, M. D. (1998). Static and dynamic finite element analyses of an idealized structural model of vertebral trabecular bone. *Journal of Biomechanical Engineering, 120*(2), 267–272.

28. Nadai, A. (1950). *Torsion of a round bar. The stress–strain curve in shear. In theory of flow and fracture of solids.*

29. Gibson, L. J. (1985). The mechanical behaviour of cancellous bone. *Journal of Biomechanics, 18*(5), 317–328.

30. Bayraktar, H. H., & Keaveny, T. M. (2004). Mechanisms of uniformity of yield strains for trabecular bone. *Journal of Biomechanics, 37*(11), 1671–1678.

31. Shi, X., Wang, X., & Niebur, G. (2009). Effects of loading orientation on the morphology of the predicted yielded regions in trabecular bone. *Annals of Biomedical Engineering, 37*(2), 354–362.

32. Burr, D. B., et al. (1997). Bone microdamage and skeletal fragility in osteoporotic and stress fractures. *Journal of Bone and Mineral Research, 12*(1), 6–15.

33. Taylor, M., Cotton, J., & Zioupos, P. (2002). Finite element simulation of the fatigue behaviour of cancellous bone*. *Meccanica, 37*(4–5), 419–429.

34. Moore, T. L. A., & Gibson, L. J. (2004). Fatigue of bovine trabecular bone. *Journal of Biomechanical Engineering, 125*(6), 761–768.

Chapter 3
Methodology of Fatigue Life Simulation in Trabecular Bone

Abstract Simulation of trabecular bone starting by micro CT-scan data and using Mimics software to construct the corresponded part of structure will be included. In this chapter, explanation of choosing sample of bone for Micro CT-scan will be covered. Computer tomography produces a volume of data that can be manipulated through a process known as "Windowing" in order to demonstrate various bodily structures based on their ability to block the X-ray beam. In this chapter after reconstructing bone in the Mimic software, trabecular bone model built up in size of 10 mm as radius and 20 mm as height.

3.1 Introduction

Methodology is generally used to solve the problem and it refers to more than a simple set of methods; rather it refers to the rationale and the philosophical assumptions that underlie a particular study relative to the scientific method. In generally, there are two types of research that conducted to solve the problems, which are experimental, numerical and analytical and this study concentrates on numerical analysis.

Procedure of simulation and analysis of the trabecular bone, as well as flow chart of instruction will be covered in this chapter. Simulation of trabecular bone starting by micro CT-scan data and using Mimics software to construct the corresponded part of structure will be included. Then constructed 3D model by Mimic software will be used in the FE package software to prepare for analysis. In this part, mesh generation play an important role because it will impact on final results.

3.2 Methodology Flow Chart for Fatigue Analysis of Trabecular Bone

Methodology flow chart for this project is presented in Fig. 3.1.

© The Author(s) 2016
M. Mostakhdemin et al., *Multi-axial Fatigue of Trabecular Bone with Respect to Normal Walking*, Forensic and Medical Bioinformatics,
DOI 10.1007/978-981-287-621-8_3

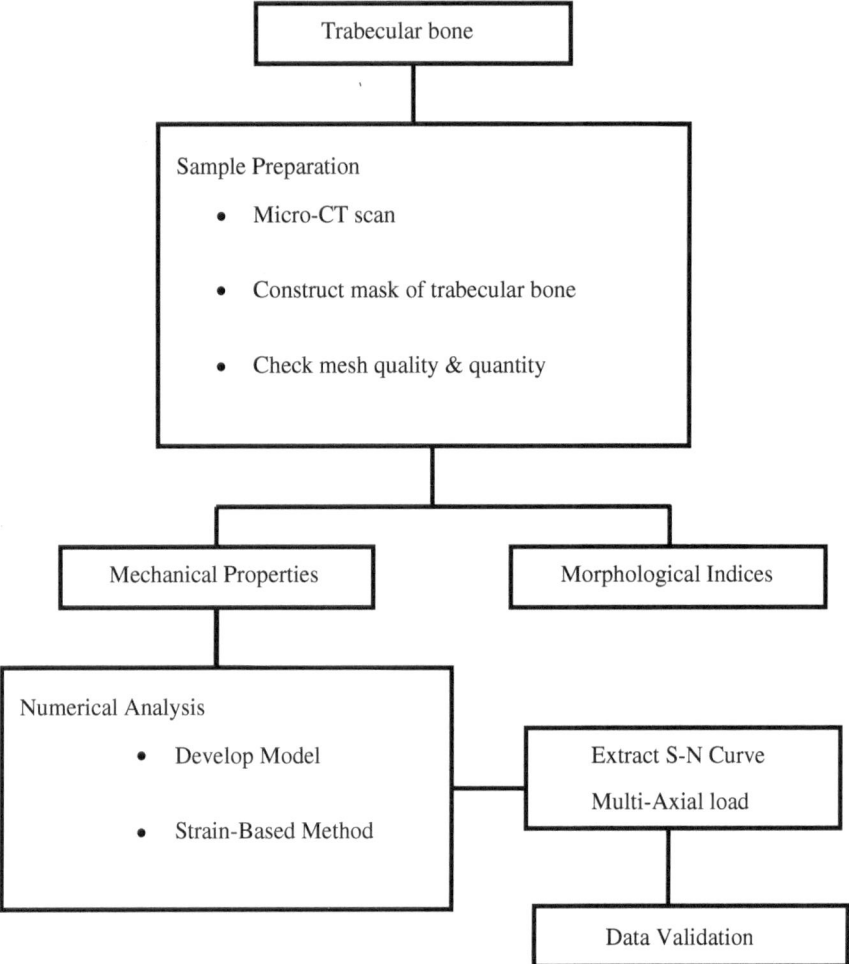

Fig. 3.1 Methodology flow chart

3.3 Material Preparation for Fatigue Simulation of Trabecular Bone

In this part of chapter, explanation of choosing sample of bone for Micro CT-scan will be covered. Since human bone for test as sample is not easy to find, typical bone, which is similar to human bone as point of mechanical properties, are necessary. Among various bone samples that would be chosen for this research, bovine bone and pig bone are more similar respect to mechanical properties. After preparation of bovine bone a piece like cylindrical with 8 mm as diameter and 20 mm as height cut and subjected to water jet for removing morrow in sample,

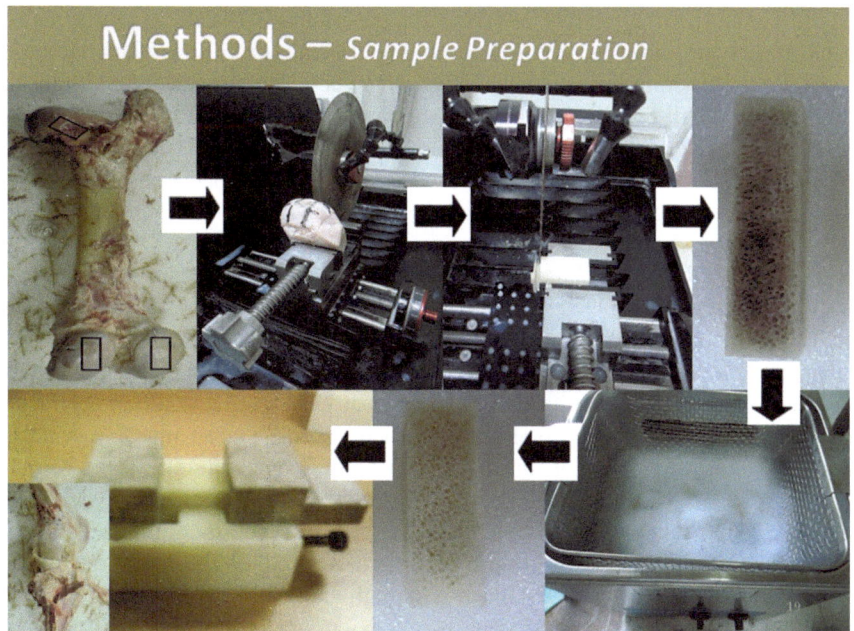

Fig. 3.2 Sample preparation of trabecular bone for analysis

specimen was kept hydrated, wrapped in the plastic, in $-20\,^{\circ}\mathrm{C}$ tight in container till mechanical testing date [1]. Figure 3.2 shows process of sample that was prepared for analysis.

3.4 Micro CT-Scan of Trabecular Sample

Computer tomography produces a volume of data that can be manipulated through a process known as "Windowing" in order to demonstrate various bodily structures based on their ability to block the X-ray beam.

The Hounsfield Units (HU): Is a linear transformation of the original linear attenuation coefficient measurement into one in which the ratio density of distilled water at standard pressure and temperature (STP) is defined as zero Hounsfield units (HU), while the ratio density of bone at STP is defined at 700–2000 HU. This value obtained by using Eq. 3.1 (x Is symbol of material used in this formula).

$$HU = 1000 * \frac{\mu_x - \mu_{water}}{\mu_{water}} \tag{3.1}$$

Micro computed tomography is useful for generating 3D scaffold structures for rapid prototyping use, as well as morphological analysis of manufactured structures. Reconstructing bone likewise physiologic connectivity density and trabecular separation will require further of processing trabecular bone in Micro CT scan free form fabrication and high resolution imaging techniques enable the creation of biomimetic structure. A 3D-CAD model of trabecular bone was produced via micro CT and export it into the other software for engineering or biomedical analysis is highly demanded [2].

Using of Micro-CT scanner (Sky Scan 1172 High Resolution micro-CT scanner) support scanning trabecular bone in the medical lab.

This is the ability of Micro-CT scan using layer by layer to capture the image taken from trabecular bone. However, all layer included in this image are made of thin layer, somehow, each of layer is less than 0.08 μm. In addition each layer stack to another one, thus, total layer of each micro-CT scan image was measured and totally 873 slice of image build our required sample up. In the Mimics software these layers stock together to construct 3D model, after this step, constructed 3D model exported to Magic software which is sub-programmed software in Mimic Materialise aimed for mesh generation on models. Then, generated mesh imported to Amira software for make it smooth and avoid gap on surface area of bone respect to generated mesh, Fig. 3.3 shows trouble with mesh by smoothing and without smoothing. Since this technique was applied, consequently, support a model to import to the FE analysis. In this study after reconstructing bone in the Mimic software, trabecular bone model built up in size of 10 mm as radius and 20 mm as height. Figure 3.4 shows the sample of trabecular model constructed in Mimics software and femoral head, which our model extracted from.

3.4.1 Reconstruction 3D Model of Trabecular Bone for Analysis

Medical imaged based software technology (Mimics Software) is software especially by materialize for medical image processing. Use Mimics for segmentation

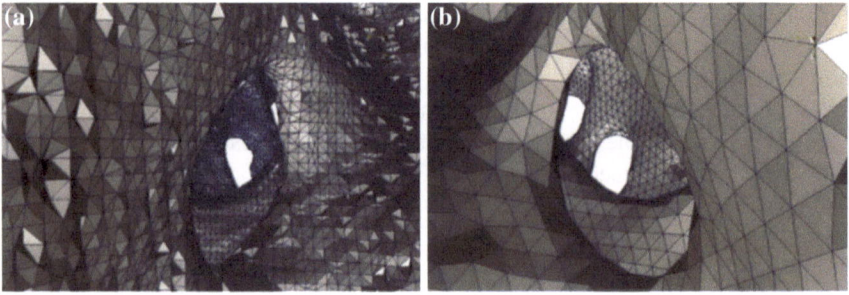

Fig. 3.3 Mesh generated on trabecular bone mask **a** without smoothing **b** with smoothing

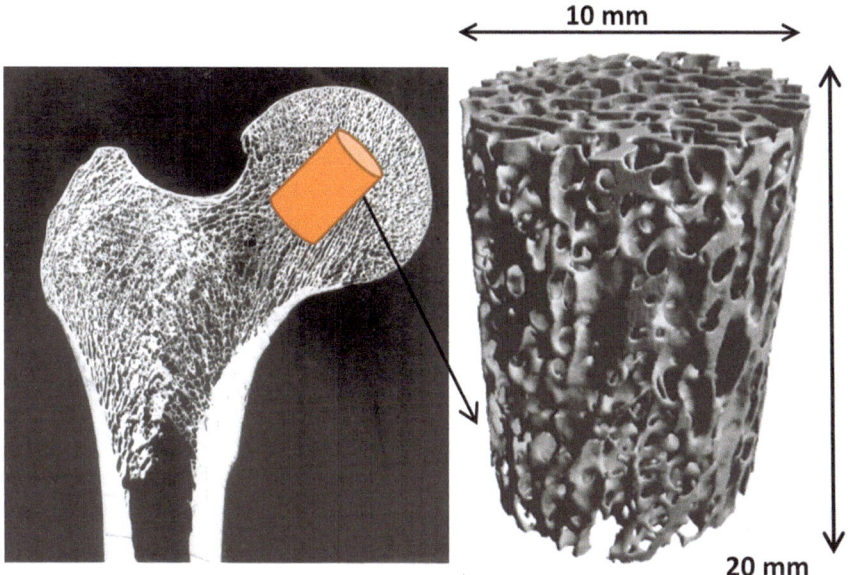

Fig. 3.4   Trabecular bone constructed in Mimics software

of 3D medical images (coming from CT MRI, Micro CT, CBCT, Ultrasound and Confocal Microscopy) and the result will be highly accurate 3D model of patient anatomy. It can then use these patient specific models for a variety of engineering application.

3-matic: is truly unique software because it is able to combine CAD tools with pre-processing (Meshing) capabilities to do this it works on triangulated (STL) files. It is extremely suitable for organic freedom 3D data such as anatomical data resulting from the segmentation of medical images (from Mimics) because of this better to call anatomical CAD. Import anatomical data in 3-matic to experience real engineering on anatomy with 3-matic we can conduct trough 3D measurement and analysis design and implant or surgical guide or prepare a mesh for finite element modelling.

Materializes biomedical software engineering software solution Mimics innovation make it possible to import 2D image data and transform them to 3D image data which helps to remodel inaccessibility of the body be easily overcome using non-inversing image technique and materializes 3D software solution.

The innovation Mimics suits is combined by two powerful software solution Mimics and 3-matic beside process and model image data Mimic also design, mesh and perform advanced analysis the seamless link between Mimics and 3-matic streamline your workflow and gives you flexibility. It bridges the gap between medical image data and the variety application such as FEA, additive manufacturing and implant design.

3.5 Process of Finite Element Analysis

Process of investigation of mechanical properties and hard tissue properties of tra-becular bone by using experimental methods need labour intensive and spending long time for testing, even, high-tech test machines that use extensometer for dis-placement within samples is not exist to measure displacement initiated by tor-sion and axial compression. It means that considering of contribution of each type of load on trabecular stress and strain distribution by experimental test would be in trouble. In contrast, numerical model development can simulate any combina-tions of loads and identify portion of each load on results separately. The ability to determine trabecular bone tissue elastic and failure properties by numerical analy-sis is highly demanded. Numerical analysis for complex structures solve problem by discretizing model and gives the approximate solution, in addition this solu-tion will be closed to accurate solution by some consideration in setting model up such as increase relative tolerance of solver during solving. Furthermore, this set up just is not limited to the tolerance, also might be related to mesh quality and quantity (shape function, order of nodes, shape of element), boundary condition respect to define fixed part or rotatable part and quantity of surface of fixed bound-ary that effect on computation time due to larger or smaller of stiffness matrix of model, even consideration of some technique such as composite face which ignore broken edge or irregular surfaces that affect on mesh generation and consequently improving in result and time of solution.

3.5.1 Introduction to Finite Element Modelling

Finite element method is a numerical method that can be used for solving engi-neering problems. It is also applicable more for complex geometries combined with loading and various boundary conditions in the sense of different physics analysis such as stress, fatigue, fluid and thermal analysis. In this study, finite element method is applied to analyse the fatigue behaviour of trabecular bone with different morphologic indices. Fatigue evaluation might be considered by analytical method, which is most useful in 2D analysis of trabecular structures, and numerical analysis for 3D modelling. In numerical analysis, first of all stress analysis must be done for model, then fatigue evaluation based three different methods which are stress-based, strain-based and crack-growth model will take a place. In the following step, procedure of numerical analysis will be described completely.

3.5.2 Geometry of Trabecular Bone in Finite Element Packages

Geometry of trabecular bone is counted as complex structure, so certain setting up must be considered to construct and import in FE software. Depend on what structure and quality is needed for analysis two types of parameters must be manipulated.

3.5.2.1 Discretization of Trabecular Bone

In this study Mimic software has been used to reconstruct trabecular bone after, building 3D model up, export model to Magic software which is embedded as subprogram in Mimic to setting some parameters such as surface triangle, quality of mesh, and angle of triangles and so on. Changing in the model in this part might play an important role in the finite element analysis and its results. In fact changing parameters in reconstruction bone software (Mimic) that is effect on mesh quality parameters is counted as a function of mesh quality in FE package, based on following parameters mesh and bone surface might be change. A perfect mesh would consist of triangles that are all equilateral; it means that each side of triangle must have same length and 60° as its angle.

1. Quality parameters (R-in/R-out, Height and Base, Skewness, Smallest angle, Largest angle, Angle ratio, Edge ratio, Equi-angle skewness, Stretch)
2. Inspection parameters (Largest angle, Height, Smallest edge length, Largest edge length, Bad Edge, Surface smoothness, Non-manifold edge, Peak, Shaft, Sharp geometry, Marked)
3. Growth parameters

To have high quality of created triangles in Magic, every parameter mentioned in first and second procedure must be inspected to show low quality triangle, Fig. 3.5 shows a Height/Base quality histogram. After clearance of low quality of triangles, critical triangle was erased and mesh manipulated or re-generated until acceptable quality appear and inspection procedure would be re-check to be sure all elements satisfy required standard of size and dimensions.

In Fig. 3.4 low quality triangles are appeared as red or orange colour, however, at the right side of model, there is a quality histogram slider, which might be changed to define range of low and high quality band. After checking each parameter, low quality triangle must be manipulated or removed or discretized to smaller triangle to improve quality of mesh and surfaces.

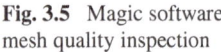 **Fig. 3.5** Magic software mesh quality inspection

To mesh generation of the trabecular bone in the Mimic software, various methods has been presented that is sub-algorithm of meshing part

1. Split-Based
2. Growth factor
3. Optimized-Based

Any of this method can control especial factors in the mesh generation. Split-Based methods can control maximum quality threshold. Once the split-based mesh is done the growth of triangles might be too big that might cause problem when generating 3D mesh from the surface mesh, to smooth the transition from small triangles to large triangles re-mesh based on growth factor can be used. In Split-Based method, quality threshold must be 0.5 to satisfy acceptable quality of mesh.

Since Magic generate mesh on 3D mask and inspections on mesh quality and quantity might has a useful contribution on finite element pre-processing step, but the size and number of elements in FE package make the most important part of solution. Since bone surface include many peaks and valley and sharp points in

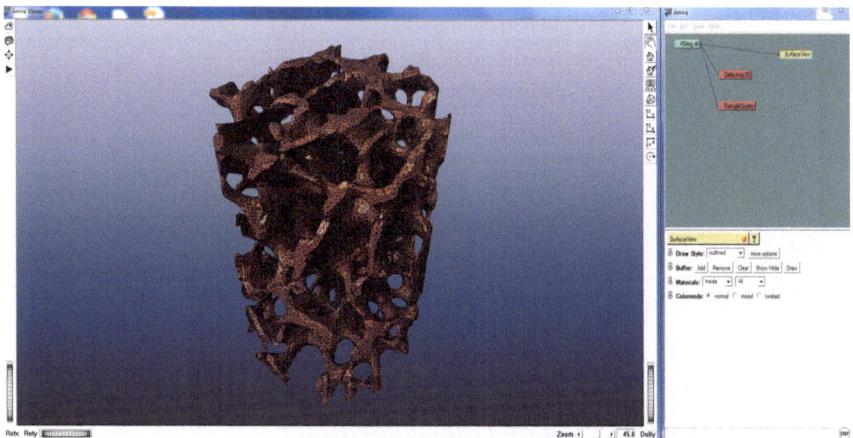

Fig. 3.6 Smoothing surface of trabecular bone in Amira software

pre-processing part of FEA, imported file would be recognized by all this irregularities on surface, so FE pre-process solver consider all edges and vertices and fit mesh based on these irregularities. Although increase number of element, improves accuracy of results. The point is many number of element make bigger size of stiffness matrix and effect on increment time of solution as well as face with lack of memory as on of the error during solution processing. To solve such a problem, Amira software applied to smooth surface after reconstruction of mask in magic software. In Amira, algorithm was developed to smooth irregularities of surface and consequently mesh generation part of FE package will be able to generate mesh regularly. Figure 3.6 shows our model in Amira software and its algorithm to smooth surface.

3.5.3 Import Parameters for Finite Element Package Respect to Trabecular Bone Samples

After constructing bone in Mimics, import 3D model into finite element package need to set up some parameters, that are make a useful contribution in transferring data to FE package. All these parameters were set up after analysing of imported model respect to the mesh quality and quantity.

The point is, changing in result might be expected by changing these parameters, and however, after changing parameters and analysing mesh quality technique, it has been found that these parameters are not function of mesh quality and just affect the surface regularity or irregularity. Table 3.1 shows all import parameters into COMSOL Multi Physics software.

Table 3.1 Imports parameters into COMSOL software

Import parameters	Values
Maximum angle (°)	110
Maximum boundary neighbour	5
Maximum relative area	0.0003
Maximum neighbour angle	0.6
Removal of small boundary	0.1
Maximum angle to extend	0.6
Detect extrude boundaries	0.03
Maximum curvature deviation	5 or 3
Detect constant	1

3.5.4 Mesh Information of Trabecular Bone in Finite Element Process

Because of irregularity of geometry, tetrahedral element must be applied to cover irregular surfaces completely. Table 3.2 is mesh characteristic for different samples has been analysed.

Mesh quality technique is one of the most applicable tools in COMSOL software that extract variety of colour for element based on their quality. Figure 3.7 shows mesh quality of tetrahedral element has been applied sample.

Vertical model of trabecular bone that shown high quality of element, as illustrated majority of element have red colour (value 1) that means high quality.

3.5.5 Fatigue Model Categories

Fatigue modelling is categorized into three models

1. Stress-Based Method
2. Strain-Based Method
3. Fatigue crack Method

A schematic of fatigue analysis approaches is as illustrated in Fig. 3.8.

Table 3.2 Mesh characteristic and information of samples

Sample	Number of elements	Tetrahedral element	Triangular element	Average element quality	Average growth rate	Maximum growth rate	Mesh volume (mm³)
Vertical model	270,043	270,043	111,880	0.6794	1.792	15.69	85.74

Fig. 3.7 Mesh quality of vertical model

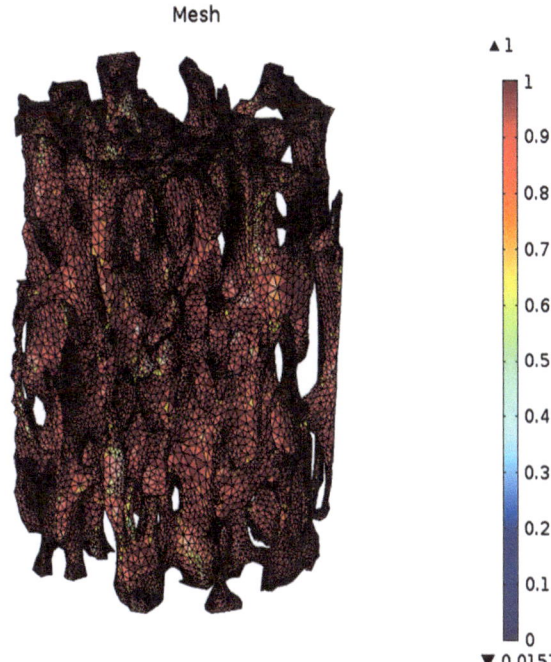

Mesh

3.5.5.1 Stress-Based Method in Fatigue Analysis

Stress-Based approach is useful when the range of stress is not high and due to these reason long lives (*more than* 10^4 *cycles*) for components will be expected. In addition, stresses will not exceed than yield strength so and elastic region.

3.5.5.2 Strain-Based Method in Fatigue Analysis

Strain-Based approach is applicable for components faced with high stress amplitude and this reason yield plastic region growth in low cycles consequently fatigue life will be decreased. In this analysis, strain-based method has been applied for trabecular bone because include of many notches, furthermore, stress localization due to stress concentration factor will be expected. In fact in such problems strain can be measured and this quantity would be excellent quantity for correlating for low cycle fatigue. The local strains can be well above the yield strains in such problems and it would be difficult to measure stresses rather than measuring strain. In notched components subjected to cyclic loading, the behaviour of material at the root of notch is best considered in terms of strain. As long as there is elastic region in neighbour of plastic zone, calculation of the strain are easier than stress. Following graph represent low cycle and high cycle fatigue.

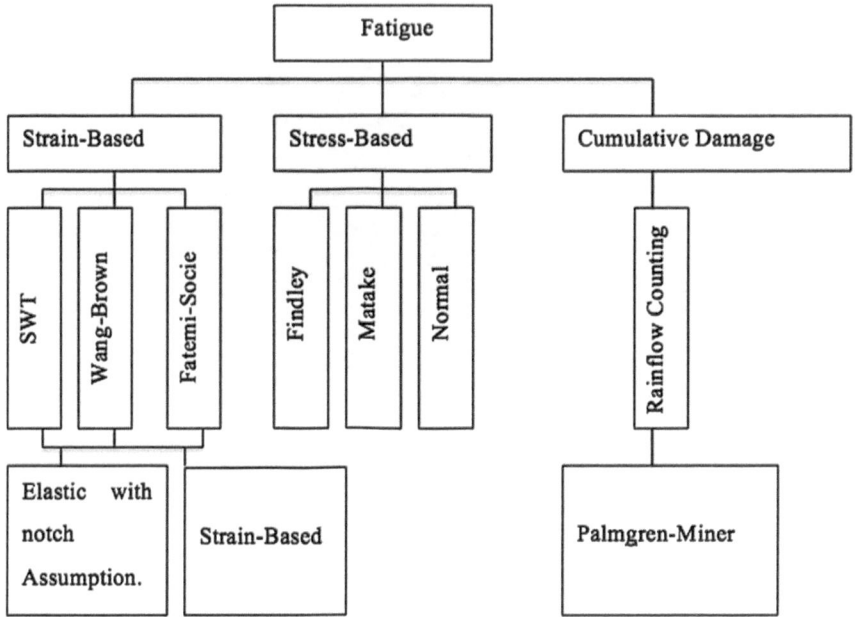

Fig. 3.8 Fatigue analysis approaches

For low cycle fatigue, strain-based approach use elastic strain and plastic strain as following equation

$$\varepsilon_T = \varepsilon_e + \varepsilon_p \tag{3.2}$$

Figure 3.9 illustrates strain-life curves, showing total elastic, and plastic strain components [3].

In this section foundation of fatigue analysis procedures and its points consideration will be explained briefly. Prediction of fatigue behaviour of components

Fig. 3.9 S-N curve of strain-based method [4]

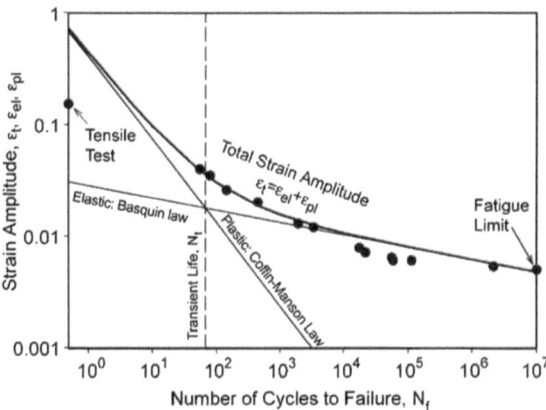

will be possible by monotonic tension test, which will be accessible by strain controlled fatigue behaviour. Before discussing on method to find how fatigue data will be extracted, material stress-strain response is covered first.

Bone tension test is such as other material tension test; by considering specimen of bone and gage length on it engineering stress-strain curves will be obtained as S that indicated stress by following equation.

$$S = \frac{P}{A_0} \tag{3.3}$$

where P is the axial force and A_0 is original cross section area. True stress, σ, in this test also will be calculated as

$$\sigma = \frac{P}{A} \tag{3.4}$$

where A is the instantaneous cross-sectional area, and true strain is obtained by

$$\varepsilon = \ln(\frac{l}{l_0}) \tag{3.5}$$

For small values of strain about 2 % true stress and engineering stress-strain are the same. However, with inelastic behaviour, we can assume constant volume due to this reason that plastic strain will not affect on volume change. Then the following relationships will be extracted.

$$\sigma = S(1 + e) \tag{3.6}$$

$$\varepsilon = \ln(1 + e) \tag{3.7}$$

These equations are valid up to necking point (ultimate tensile strength), but after this point because of localized plastic strain, strain will not be uniform. Total strain is where elastic strain and plastic strain will be added as follows

$$\varepsilon_T = \varepsilon_e + \varepsilon_p \tag{3.8}$$

and plastic strain in log-log coordinates plot is drawn by

$$\sigma = K(\varepsilon_p)^n \tag{3.9}$$

where K is the strength coefficient and n is strain-hardening exponent (slope of the line), and then the total strain equation is

$$\varepsilon_T = \varepsilon_e + \varepsilon_p = \frac{\sigma}{E} + (\frac{\sigma}{K})^{1/n} \tag{3.10}$$

This relationship called "Ramberg-Osgood relationship"; plastic deformation is counted as a critical parameter in fatigue process due to crack nucleation from this region. Since of plastic deformation plays an important role in fatigue process, therefore, cyclic-strain controlled test provide excellent condition to tracking fatigue behaviour. Consequently, we utilize this method for fatigue data of trabecular bone (Fig. 3.10).

Fig. 3.10 Stress-strain of trabecular bone

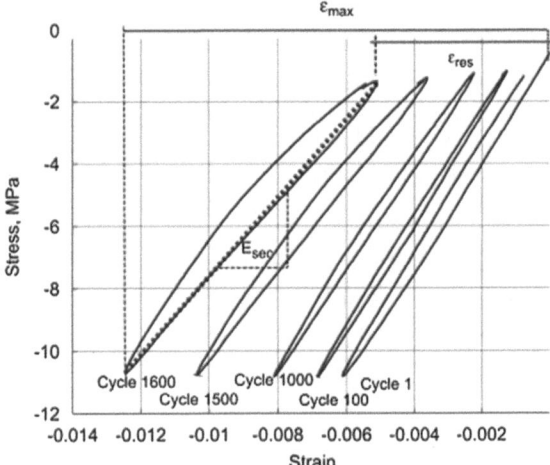

As it is cleared in strain-life curves, plastic strain will be considered at short live and large strain, in contrast, elastic strain will be predominant at low strain but in longer lives. Based on this graph, intercept of two lines (plastic strain and elastic strain) is $\frac{\sigma_{f'}}{E}$ for elastic part and ε'_f for plastic part. And their slope is introduced as b for elastic line and c for plastic line respectively. This data yield following equation called, strain-life data.

$$\frac{\Delta\varepsilon}{2} = \varepsilon_a = \frac{\Delta\varepsilon_e}{2} + \frac{\Delta\varepsilon_p}{2} = \frac{\sigma_{f'}}{E}(2N_f)^b + \varepsilon'_f(2N_f)^c \qquad (3.11)$$

All terms of strain-life data is introduced as follow

$\frac{\Delta\varepsilon}{2}$	Total strain amplitude
$\frac{\Delta\varepsilon_e}{2}$	Elastic strain amplitude
$\frac{\Delta\varepsilon_p}{2}$	Plastic strain amplitude
ε'_f	Fatigue ductility coefficient
c	Fatigue ductility exponent
$\sigma_{f'}$	Fatigue strength coefficient
b	Fatigue strength exponent
E	Modulus of elasticity
$\Delta\sigma/2$	Stress amplitude

The straight-line elastic part can be changed to following equation

$$\frac{\Delta\sigma}{2} = \sigma_a = \sigma_{f'}(2N_f)^b \qquad (3.12)$$

This equation known as Basquin's equation, in addition,

$$\frac{\Delta \varepsilon_e}{2} = \varepsilon_f'(2N_f)^c \tag{3.13}$$

Which known as Coffin-Manson relationship, now the point is, intersection of elastic line and plastic line which is called transition fatigue life can be obtained by following equation

$$2N_t = (\frac{\varepsilon_f' E}{\sigma_{f'}})^{\frac{1}{b-c}} \tag{3.14}$$

For lives less than $2N_t$ the deformation is counted as mainly plastic, however, for longer data larger that $2N_t$ the deformation is mainly elastic.

The strain-based approach support high-cycle and low-cycle fatigue, this approach also support long-life processes that small plastic strain would be presented. In this case, plastic strain from strain-life data will be removed and just Bassquin's equation will be remained. Consequently, strain-based approach is widely applicable in low-cycle and high-cycle components.

After fitting data to obtain all strain-life properties, these properties, stress and plastic strain are counted as independent variables and fatigue life as dependent variable as shown mathematically below.

$$\text{fatigue life} = f(\sigma_{f'}, \varepsilon_f', b, c, \text{stress}, \text{plastic strain})$$

This equation is satisfied because fatigue life cannot be controlled and depends on applied strain amplitude.

3.5.6   Material Properties of Trabecular Bone

3.5.6.1   Elastic-Plastic Properties of Trabecular Bone

In the various study elastic-plastic properties has been investigated, different mechanical properties depend on various anatomic sites and its microstructure, elastic and yield properties of trabecular tissue are similar to cortical one [5]. After analysis in tensile yield strain computationally and experimentally, it has been found that tensile yield strain are uniform across site and species. In-elastic region that has been measured experimentally show smooth transition from the yield to the ultimate point that with standard bi-linear model will not be modelled. In this study Voce model data has been that independent variable is plastic strain, while the dependent variable was described by means of 3 parameters, the yield stress σ_y, the post yield hardening stress R and the exponential hardening exponent B [6] (Fig. 3.11).

Fig. 3.11 Standard stress-strain curve according to Voce model, with the corresponding parameters, E (Elastic modulus, GPa), σ_y (yield stress, MPa), ε_y (yield strain,-), σ_U (Ultimate stress, MPa), ε_{pl} (plastic strain,-), B (exponential strain hardening coefficients,-), R (post-yield hardening stress, MPa)

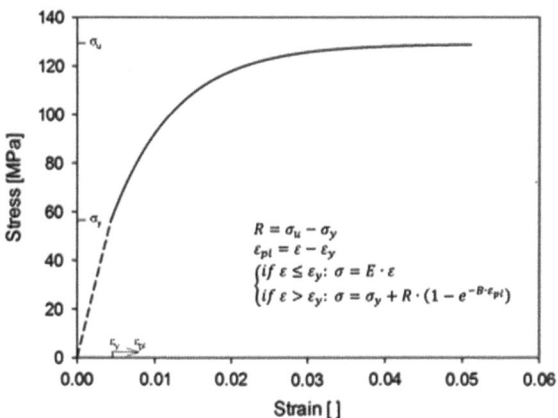

Table 3.3 Fatigue S-N curve data for trabecular bone

Fatigue parameters	Symbol	Value
Initial yield stress (MPa)	σ_y	8.4
Kinematic tangent modulus (GPa)	E_k	0.9
Fatigue ductility coefficient	ε_f'	0.352
Fatigue ductility exponent	b	−0.981
Fatigue strength coefficient (MPa)	σ_f'	6
Fatigue strength exponent	c	−0.096

3.5.6.2 Fatigue S-N Curve Data for Trabecular Bone

Since the fatigue analysis of trabecular bone in strain-based model needs its coefficients, consequently all these coefficients come from experimental part as defined in Table 3.3.

3.5.6.3 Isotropic and Kinematic Hardening Model

In cyclic loading condition, means loading and unloading, deformation of structure plastically and unload it then load it again which cause increase in plastic flow and material resistance to reach plastic region at previous point. Consequently, yield point or elastic limit increase that this phenomenon defined as strain hardening. Two model of strain hardening is introduced as follows

- Isotropic hardening
- Kinematic hardening

In isotropic hardening if increase the load then unload it and reload it again, yield stress of material will be increased as compared with preceding point, this steps continuous till structure will deform elastically. Furthermore, isotropic hardening means that if the yield stress increase in tension, compression yield stress also will be increased even though you might not have been loading the specimen in compression.

Since Bauschinegr effect is not consider in isotropic hardening and predict that material will be hardened after few cycles till elastic response is satisfied, this type of hardening is not suitable for specimens under cyclic loading. Consequently, kinematic hardening is appropriate for this part and the material will be softens in compression, so it is quite good enough to use for material under cyclic behaviour and Bauschinger effect.

Strain hardening model is developed for trabecular bone to describe yielding of trabecular bone at the continuum level. To develop plasticity-like model in strain space for a yield envelope expressed in terms of principal strain having asymmetric yield behaviour. So far constitutive model for non-linear trabecular bone has been developed and based on it, plasticity-like model capture expected behaviour with the kinematic and isotropic hardening as well as tension compression yield strength as illustrated in figure below [7] (Fig. 3.12).

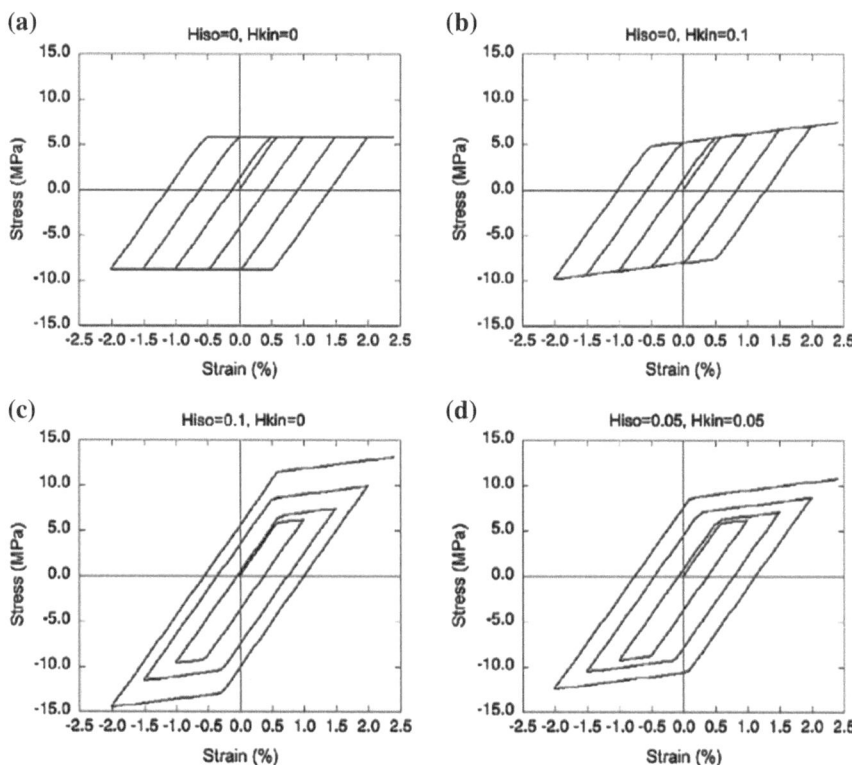

Fig. 3.12 The stress–strain behaviour at one of the integration points of an eight-node brick subjected to cyclic loading, four test cases were considered assuming material behaviour as **a** elastic perfectly plastic; **b** kinematic hardening 0.1 times the elastic modulus; **c** isotropic hardening 0.1 times the elastic modulus; and **d** both kinematic and isotropic-hardening

3.5.7 Boundary Condition of Trabecular Bone for Mechanical Analysis

Since material properties and mesh sequence have been set up for analysis, boundary condition and loading are introduced in this section. A sample imported as model into FE package software need to identify boundary condition and load area on structure. First set of analysis has been done for multi-axial loading, fixation part of trabecular bone is as shown below in Fig. 3.13.

Lowest surface is fixed in all direction and load was imposed on upper surface.

$$U_x, U_y = 0 \tag{3.15}$$

$$R_x, R_y = 0, R_z \neq 0 \tag{3.16}$$

3.5.8 Load Amplitude Imposed on Hip Based on Normal Walking

In this study because the loading is in three different directions X, Y and Z based on walking with sensors that plot three curves in the system and this curves are

Fig. 3.13 Load and boundary on trabecular bone

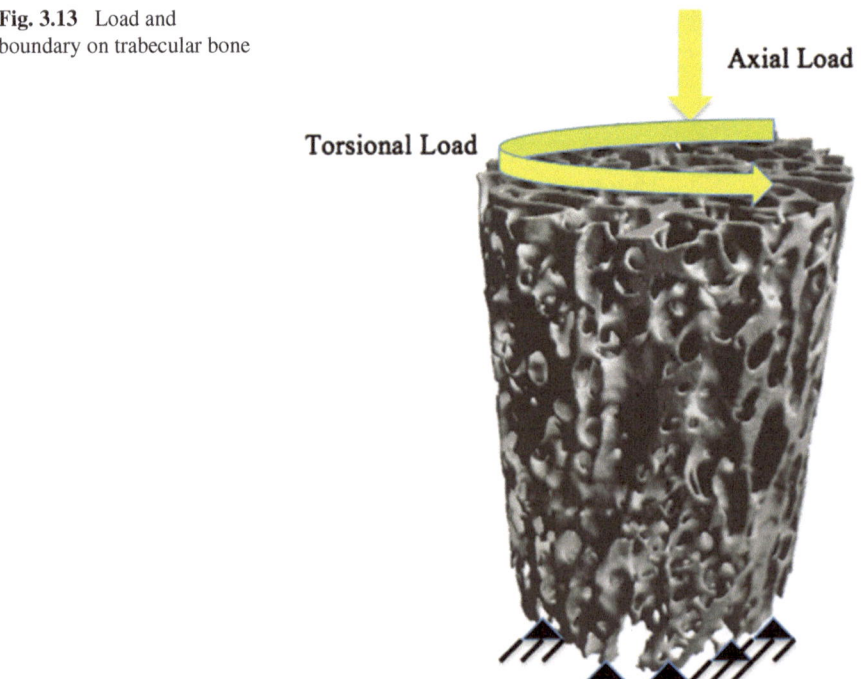

Axial Load

Torsional Load

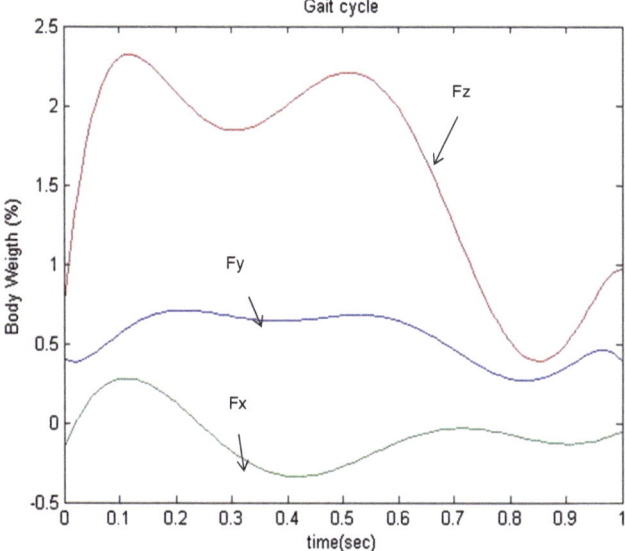

Fig. 3.14 Curve of normal walking loads trends by Matlab software

plotted in Fig. 3.14. Then two of curves (F_x and F_y) was used for torsional load and F_z was used for axial load. Furthermore in torsional load, fx and fy multiplied with 5 mm as radius of sample present moment force imposed on sample.

Continuous loading indicate non-zero stress or negative stress, since it is clear in gait loading graph all curves has no intercept with x axis or negative value (below x axis), and axial loading can be extracted from these curves. Since the gait loading graph is represented body weight percentage, it is recommended that applying average body weigh among adults to find its best fit polynomial equation as loading equation simulation.

Finally, graph of body weight in 60 kg is selected, in addition, to have a better curve fitting on the gait cycles, highest orders were employed to get the best representative of curves. However, the contra of using high order polynomials is the curve is starting to scatter at both tips of the trends. Furthermore, heuristic method on the nearest polynomial equations is getting more complicated to match. Responding to the software requirement, loadings in parametric study need a special equation to represent as loadings in the calculation of element matrixes.

3.5.8.1 Polynomial Equation of Gait Loading for Trabecular Bone

As necessary of loading equation to the COMSOL Software, original trend of gait cycle on Fig. 3.14 must translate into equation as represent gait loading. There 3 axes loading exerted form one completed gait cycle. In approaching those 3 equations, by using Matlab function, commanding the through code to extract as nearest polynomial function as gait cycle presenting in Fig. 3.14

From a simple coding, Fig. 3.4 above pointing on the graph of polynomial extracted through Matlab software. Since the genuine graph of gait cycle trend is in body weight percentages, thus, the graph obtained by person who has 60 kg weight. Below are the polynomial equations involved in gait loading analysis.

List of polynomial for 60 kg;

$$F_x = 1e^6(-0.3091t^7 + 1.058t^6 - 1.41t^5 + 0.9215t^4 - 0.3032t^3 + 0.0441t^2 \\ - 0.0012t + 0.0002)$$

(3.17)

$$F_y = 1e^5(-0.9961t^7 + 3.4936t^6 - 4.4099t^5 + 2.181t^4 - 0.0717t^3 - 0.2461t^2 \\ + 0.0488t - 0.0009)$$

(3.18)

$$F_z = 1e^5(-2.6105t^7 + 6.3501t^6 - 3.2159t^5 - 3.2341t^4 + 3.9678t^3 - 1.4579t^2 \\ + 0.202t - 0.0043)$$

(3.19)

3.5.9 Morphological Indices Categories in Bone

Since analysis in trabecular bone is done in the direction which load is imposed on bone longitudinally, and imposed load transversally and obliquely is beyond the scope of this study. Capturing morphology characteristics of model is highly recommended to find correlation between bone morphology and its mechanical properties or failure properties.

In this study, Image J software is applied to extract morphology indices of models such as BV/TV and BS/TS.

3.5.9.1 Morphological Indices of Three Oriented Sample of Trabecular Bone

Among various morphological indices in the morphology analysis, two parameters play an important role in fatigue life of trabecular bone that is BV/TV and BS/TS. BV/TV is the bone volume over total volume that for every samples volume of bone is calculated and divided by volume of the exact dimension of analysed bone in aspect of cylindrical sample. The other parameter that is counted as crucial factor in fatigue life is BS/TS that is known as bone surface over total surface, for this sample surface area of samples calculated and then divided by surface area of cylindrical-shape sample to obtain this factor. Morphology indices is presented in Table 3.4 is belong to load imposed longitudinally as vertical sample others just represented to show anatomic variation in morphologies.

Table 3.4 Bone morphology indices for three samples	Samples	BV/TV	BS/TS
	Vertical	0.159	2.957
	45°	0.199	3.5
	Horizontal	0.172	3.042

3.6 Summary of Methodology Trabecular Bone Analysis

This chapter had an overview of the procedure of project and explanation about the method will be used in the following chapter. First of this chapter was started with flow chart which introduced process of project briefly, then explanation about material preparation for 3D scanning and import it into the Mimic software for reconstruct the model was covered. Afterward, export model and import it into FE package software was described for analysing and getting the results by its numerical method. At the bottom of line bone morphology indices for each sample was calculated and introduced.

References

1. Bayraktar, H. H., & Keaveny, T. M. (2004). Mechanisms of uniformity of yield strains for trabecular bone. *Journal of Biomechanics, 37*(11), 1671–1678.
2. Tellis, B. C., et al. (2008). Trabecular scaffolds created using micro CT guided fused deposition modeling. *Materials Science and Engineering C, 28*(1), 171–178.
3. Ralph, I., Stephen, A. F., Stephen, R. R., Fuches, H. O. (1980). *Metal Fatigue in Engineering* (2nd ed.). USA: A Wiley Interscience Publication.
4. Taylor, M., Cotton, J., & Zioupos, P. (2002). Finite element simulation of the fatigue behaviour of cancellous bone*. *Meccanica, 37*(4–5), 419–429.
5. Morgan, E. F., et al. (2004). Contribution of inter-site variations in architecture to trabecular bone apparent yield strains. *Journal of Biomechanics, 37*(9), 1413–1420.
6. Carretta, R., et al. (2013). Novel method to analyze post-yield mechanical properties at trabecular bone tissue level. *Journal of the Mechanical Behavior of Biomedical Materials, 20*, 6–18.
7. Gupta, A., et al. (2007). Constitutive modeling and algorithmic implementation of a plasticity-like model for trabecular bone structures. *Computational Mechanics, 40*(1), 61–72.

Chapter 4
Result and Discussion of Static and Dynamic Analysis of Trabecular Bone

Abstract Stress-strain analysis in multi-axial loading is considered then fatigue life prediction of trabecular bone is represented. Vertical model subjected to axial load is faced with lower stress amplitude and effective plastic strain as well. Effective plastic strain in trabecular bone is known as a crucial data for tracking crack nucleation point and damage behaviour under different types of load. Since Combination of axial and torsional load on trabecular bone make life of bone shorter than imposing each load separately, this part of study consider the trend of stress and strain within the structure. Volumetric strain and effective plastic strain data were combined to extract data for strain amplitude axis and number of cycles to failure data computed numerically used as number of cycles to failure for x-axis in four load amplitude. Since trabecular bone is not withstand against load as structure subjected to axial load can, so it would be expected that fatigue life in torsional load is decrease drastically in compare with axial load. In static analysis trabecular bone was subjected to axial load two parameter monitored which were stress versus volumetric strain and effective plastic strain that is the crucial factor of damage initiation within structure. Results show that trabecular bone subjected to axial load has considerable fatigue life which in 40 % of total load it reaches to 1226.80 cycles approximately, however, this value for torsional load reach to 42 cycles and for multi-axial 38 cycles. Torsional load known as the most effective load, which causes damage within struts and trabeculae.

4.1 Introduction

In this chapter, analysis results that have been completed will be presented. Two different types of result are covered, first stress-strain analysis in multi-axial loading is considered then fatigue life prediction of trabecular bone is represented and results being compared with the same sample but with separate imposed axial and torsional load in previous results. Finally, correlation among different imposed

© The Author(s) 2016   37
M. Mostakhdemin et al., *Multi-axial Fatigue of Trabecular Bone with Respect to Normal Walking*, Forensic and Medical Bioinformatics,
DOI 10.1007/978-981-287-621-8_4

Table 4.1 Stress and volumetric elastic strain in axial load

Load (%)	10	20	30	40
Stress (MPa)	1.58	3.16	4.11	4.56
Volumetric strain	3.09e-5	6.18e-5	9.29e-5	1.26e-4

load condition on sample will be discussed. Stress analysis first must be completed (study 1), then fatigue analysis apply the stress solution (study 1) to solve for fatigue part as (study 2). Maximum stress and effective plastic strain is covered in this chapter as static analysis and strain amplitude versus number of cycles to failure by multi-axial load that called dynamic analysis is reported. Among different bone morphology indices, volume fraction (BV/TV) and surface density (BS/BV) counted as a major factor effect on fatigue life of bone and fragility, by which this two factors calculated in this study.

4.2 Static Analysis of Trabecular Bone Subjected to Various Types of Loads

In this section four parameters will be introduced, the first stress distribution and amplitude in von misses criteria, volumetric elastic strain, effective plastic strain and elastic strain energy density. These parameters clear behaviour of the trabecular bone subjected to different load percentages in its fatigue life. Stress distribution supports study how weaken trabeculae face with cyclic load and to what extend rod-like and plate-like trabeculae have contribution on failure of structure, volumetric elastic strain use a cubic element to measure total strain in all directions, effective plastic strain clear the plastic accumulation which distributed within bone architecture and counted as the most crucial factor of crack nucleation; elastic strain energy density is known as a best factor that can indicate damage and its behaviour within structure. By use of these four parameters prediction and behaviour of fatigue life of trabecular bone would be predictable.

4.2.1 Static Analysis of Trabecular Bone

4.2.1.1 Axial Load Imposed on Trabecular Bone

Vertical model subjected to axial load is faced with lower stress amplitude and effective plastic strain as well. Here the trend of load percentage increment and its result are tabulated in Table 4.1. This table also shows how the volumetric elastic strain changes due to load increment. Volumetric elastic strain is selected due to total shrinkage of element volume under axial load in the main direction as illustrated in Fig. 4.1.

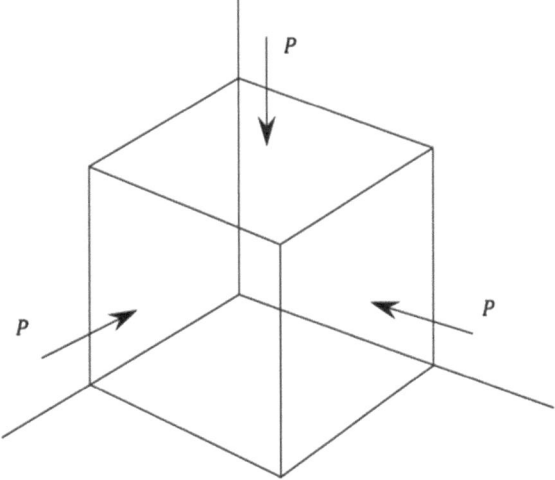

Fig. 4.1   Volumetric elastic strain in an element

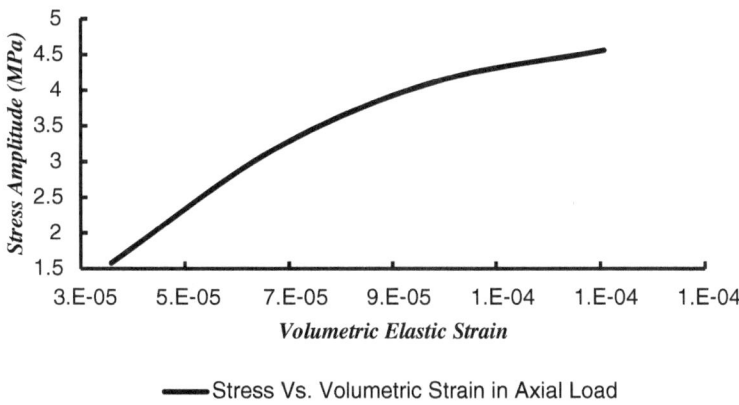

Fig. 4.2   Stress versus volumetric elastic strain in axial load

Results show that with increase of load amplitude, volumetric strain increase and this increment change smoothly after 30 % of total load. Figure 4.2 is shown trend of volumetric elastic strain by load increment.

All values of stress amplitude and effective plastic strain for axial load are tabulated in Table 4.2 obtained from numerical analysis.

Stress-effective plastic strain for vertical model I s plotted in Fig. 4.3 based on values of Table 4.1. As is clear plot start from 30 % of load.

Since vertical model included more plate-like trabeculae that make it more strength, plastic strain localized with 30 % of load, furthermore, 4.11 MPa is the stress value of this load percentage and plastic strain is 7.61e-5. Then with load

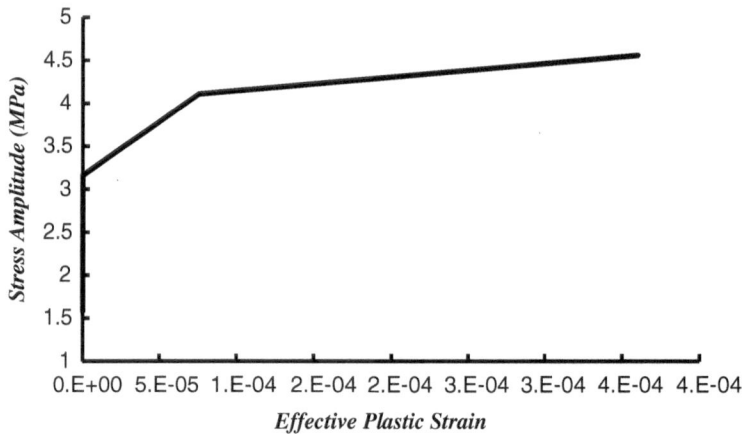

Fig. 4.3 Stress versus effective plastic strain of axial load

increment, the plastic strain has no considerable increment and it is increased with mild slope to 3.60e-4. In addition stress localization is occurred in rod-like trabeculae. In the range of 20–30 % of total load this trend increases dramatically, however, in 30–40 % this trend increase slowly. Since after 30 % of load, the changes in plastic strain can consider as effect of strain hardening of material which defined for this simulation, and increment of load and trend of plastic localization would be decreased.

4.2.1.2 Torsional Load Imposed on Trabecular Bone

Since trabecular bone that is subjected to the torsional load has not enough strength respect to axial load and struts are known as weaken tissue against torsional load, moreover, if they have been involved with osteoporosis disease fragility of struts will be more. Consequently, it is predicted that stress amplitude and plastic strain increase in this type of load respect to the axial load. Here the trend of load percentage increment and its result in volumetric strain are tabulated in Table 4.3.

In torsional load, the history is changed. The volumetric elastic strain after 20 % of total load increase drastically respect to stress and volumetric strain. As

Table 4.2 Stress and plastic strain in axial load

Load (%)	10	20	30	40
Stress (MPa)	1.58	3.16	4.11	4.56
Effective plastic strain	0	0	7.61e-5	3.60e-4

Table 4.3 Stress and volumetric elastic strain in torsional load

Load (%)	10	20	30	40
Stress (MPa)	1.54	7.98	7.88	7.97
Volumetric strain	9.10e-6	1.82e-5	8.65e-5	1.50e-4

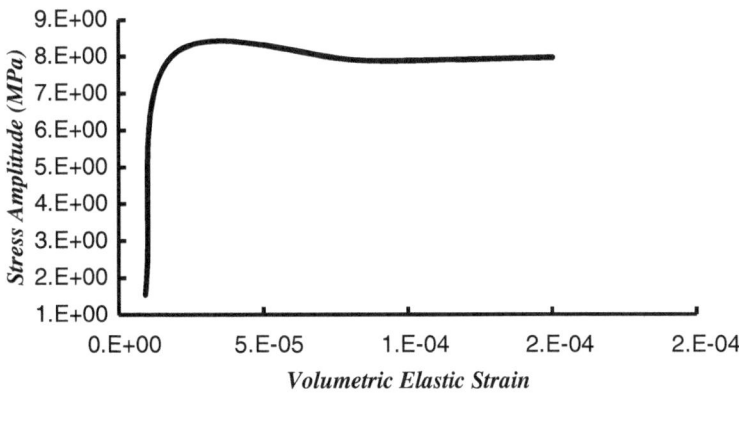

━━━Stress Vs. Volumetric Elastic Strain in Torsional Load

Fig. 4.4 Stress versus volumetric elastic strain in torsional load

is cleared in Table 4.3, after 10 % of total load, the stress amplitude in element reaches to 7.98 MPa compared with axial load that increment was 1.58 MPa. This trend show that how struts are weaken against torsional load and stress distribution over the architecture increase. Figure 4.4 present stress amplitude versus volumetric elastic strain.

Graph shows that after 30 % of total load, there is no stress and just stain increase and in 30–40 % strain is 8.65e-5–1.50e-4 respectively. In contrast, stress drastically increases in 10–20 % of load without considerable changes in elastic strain.

Effective plastic strain in trabecular bone is known as a crucial data for tracking crack nucleation point and damage behaviour under different types of load. Results tabulated in Table 4.4 show that effective plastic strain in 10 % of load is zero, however, this parameter take a place in 20 % of load with incredible values respect to axial load. Consequently, again this results show trabeculae is weak

Table 4.4 Stress and plastic strain of torsional load

Load amplitude (%)	10	20	30	40
Stress (MPa)	1.55	7.99	7.88	7.97
Effective plastic strain	0	0.00173	0.00369	0.00684

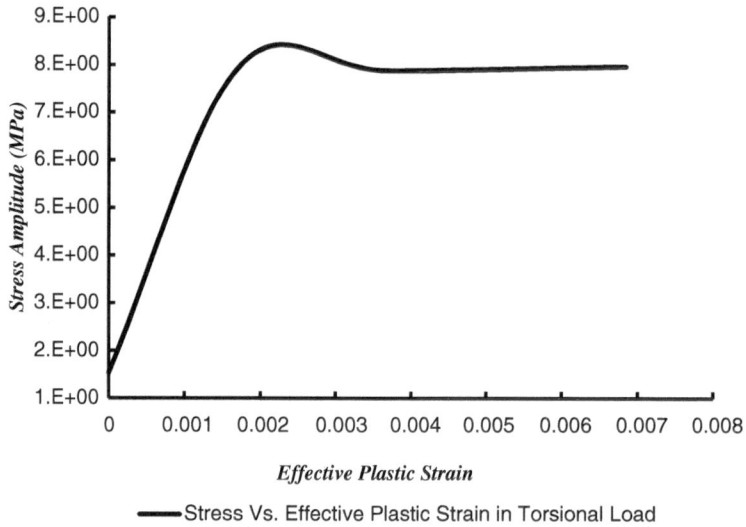

Fig. 4.5 Stress versus effective plastic strain in torsional load

against torsional load with which at 20 % of load value reach to 0.00173 and by increasing load 10 % effective plastic strain increase double as function presented below.

$$\varepsilon_{p(X_n)} \cong 0.5\varepsilon_{p(Y_n)} \qquad\qquad (4.1)$$

$$if\ Y_n = 10\,\% f_{load}(X_n) \qquad\qquad (4.2)$$

Graph illustrated in Fig. 4.5 is about Stress versus effective plastic strain in torsional load imposed on trabecular bone.

Sharply increase in stress and also plastic strain in trabecular bone subjected to torsional load is cleared within 10–20 % of load increment, then after, with the same stress, plastic strain increase and this is the point where crack initiation might occur in this region.

4.2.1.3 Multi-axial Load Imposed on Trabecular Bone

Since Combination of axial and torsional load on trabecular bone make life of bone shorter than imposing each load separately, this part of study consider the trend of stress and strain within the structure. Contribution of each load and its effect on life of trabecular bone can be considered in the sense that, stress and strain distributed all over the structure is close to which load. In this section, first volumetric elastic strain and effective plastic strain will be covered. Volumetric elastic strain versus stress results tabulated in Table 4.5 and plotted in Fig. 4.6.

According to the Table 4.5, stress amplitude within structure is too close to the torsional load, and it seems that torsional load has a main contribution on stress and strain distribution and amplitude on structure and axial load effect less on struts. Figure 4.6 show the behaviour of stress versus volumetric strain of trabecular bone in multi axial load.

Effective plastic strain plays an important role on failure of structure in the sense of plastic localization and accumulation which is counted as a factor of damage in bone. Multi-axial load must do damage on bone more than axial or torsional separately, so it will be predicted that such a load on sample cause to increase plastic strain and stress as well. Table 4.6 is presented of stress and effective plastic strain by multi-axial load on trabecular bone.

Table 4.5 Stress and volumetric elastic strain in multi-axial load

Load amplitude (%)	10	20	30	40
Stress (MPa)	1.54	7.98	7.97	8.19
Volumetric strain	9.10e-6	2.31e-5	8.10e-5	1.53e-4

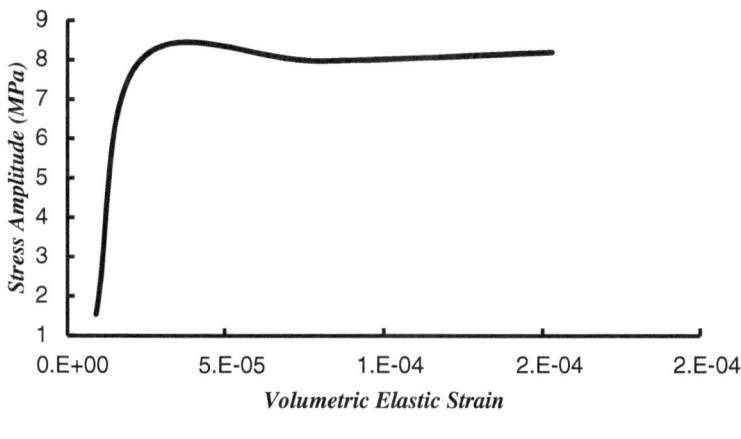

━━ Stress Vs. Volumetric Elastic Strain in Multi-Axial Load

Fig. 4.6 Stress versus volumetric strain in multi-axial load

Table 4.6 Stress and effective plastic strain in multi-axial load

Load amplitude (%)	10	20	30	40
Stress (MPa)	1.54	7.98	7.97	8.19
Effective plastic strain	0	8.66e-4	0.00511	0.00879

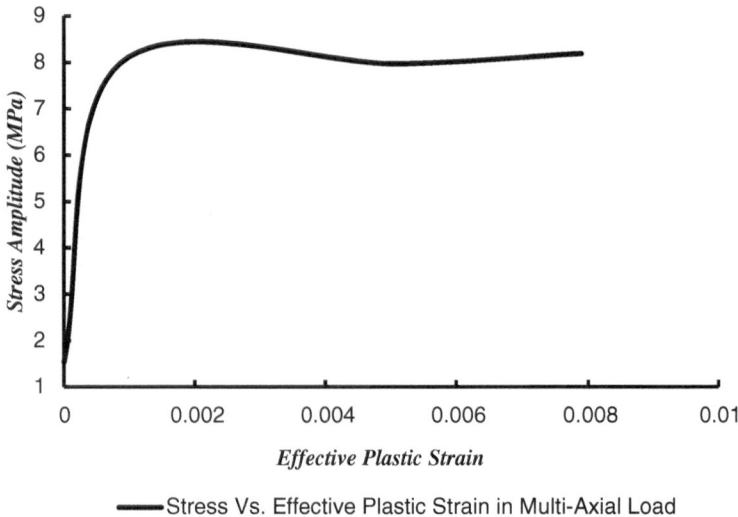

Fig. 4.7   Stress versus effective plastic strain in multi-axial load

Effective plastic strain results show that by multi-axial load, after 10 % of total load plastic strain initiated and after 20 % of total load its increment is considerable even more than torsional load. Results shows that in 20 % of load the plastic strain is less than 20 % of torsional load and this phenomenon occur because of strain hardening that after each cycle material will be more hardened and when torsional combine with axial then plastic strain initiated at 20 % with the value of 8.66e-4 compare with plastic strain in torsional that was obtained 0.00173. Figure 4.7 is about stress versus effective plastic strain by multi-axial load.

4.3 Dynamic Analysis of Trabecular Bone

In dynamic analysis, fatigue life prediction of trabecular bone is counted as the main scope of this study. Since strain-based method was applied as fatigue model, so strain amplitude versus number of cycles to failure is plotted for sample in axial, torsional and multi-axial load. In this section, volumetric strain and effective plastic strain data was combined to extract data for strain amplitude axis and number of cycles to failure data computed numerically used as number of cycles to failure for x-axis in four load amplitude.

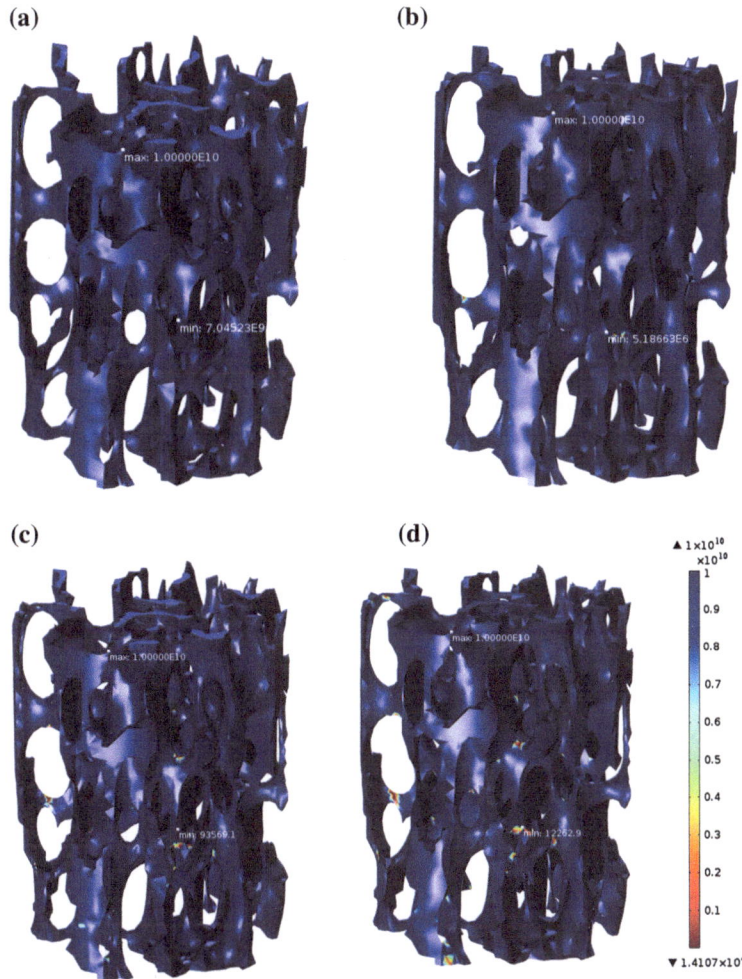

Fig. 4.8 Number of cycles to failure of axial load. **a** 10 % of total loading. **b** 20 % of total loading. **c** 30 % of total loading. **d** 40 % of total loading

4.3.1 Fatigue S-N Curve for Trabecular Bone Subjected to Axial Load

Since axial load analysis shown in the static analysis part which had no effect more than torsional load respect to plastic strain and stress distribution, so it would be expected that fatigue life of axial load is higher than other two types of load. Here fatigue analysis result of trabecular bone and its S-N curve is presented in Fig. 4.8.

In axial load, trabecular bone can withstand against load because this structure tolerate 70 % of total weight imposed on hip joint. As is clear from Fig. 4.8, in 10 % of load there is now plastic localization respect to static analysis obtained from previous section. However, fatigue life decrease when load amplitude increase and it is more clear in 20 % of load that in one of the struts plastic strain localized and then after, weaken structure faced with damage initiation by which number of cycle to failure decrease to 5.19e-6 at 20 % of total load respect to 10 % of load which N_f was 7.05e-9, this value decrease when approximately half of load amplitude imposed on bone then fatigue life drastically decrease to 1226.80 cycle. Which all this data reported in Table 4.7 (N_f means number of cycle to failure).

And plot of the axial load is illustrated in Fig. 4.9.

In the axial load, higher fatigue life is referred to 10 % of total load and as long as load amplitude increase and due to this reason stress amplitude increase and fatigue life decrease. However, porosity of structure plays an important role in fatigue life of bone. In 10 % of total loading life is 7e-9 cycles to failure and this values decrease drastically to 12262 when 40 % of total load is imposed on it. In the range of 10–20 % the fatigue life will not change too much, however, after 30 % of total load imposed, number of cycles decrease drastically to 93,569.

Table 4.7 Number of cycles to failure for axial load on trabecular bone

Load (%)	10	20	30	40
N_f	7.05e-9	5.19e-6	93569.13	1226.80

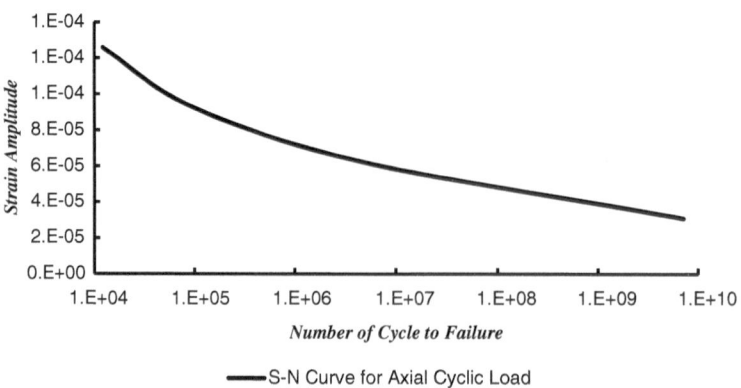

Fig. 4.9 S-N curve for axial load on trabecular bone

4.3.2 Fatigue S-N Curve of Trabecular Bone Subjected to Torsional Load

Since trabecular bone is not withstand against load as structure subjected to axial load can, so it would be expected that fatigue life in torsional load is decrease drastically in compare with axial load. Figure 4.10 is results of fatigue life prediction of trabecular bone subjected to torsional load.

In torsional load, number of cycles to failure has completely different history respect to axial load. As mention in Chap. 2 struts and trabeculaes are not withstand against torsional load and the most literature that carried out their research on this area found this problem. Furthermore Fig. 4.10 is completely showing the

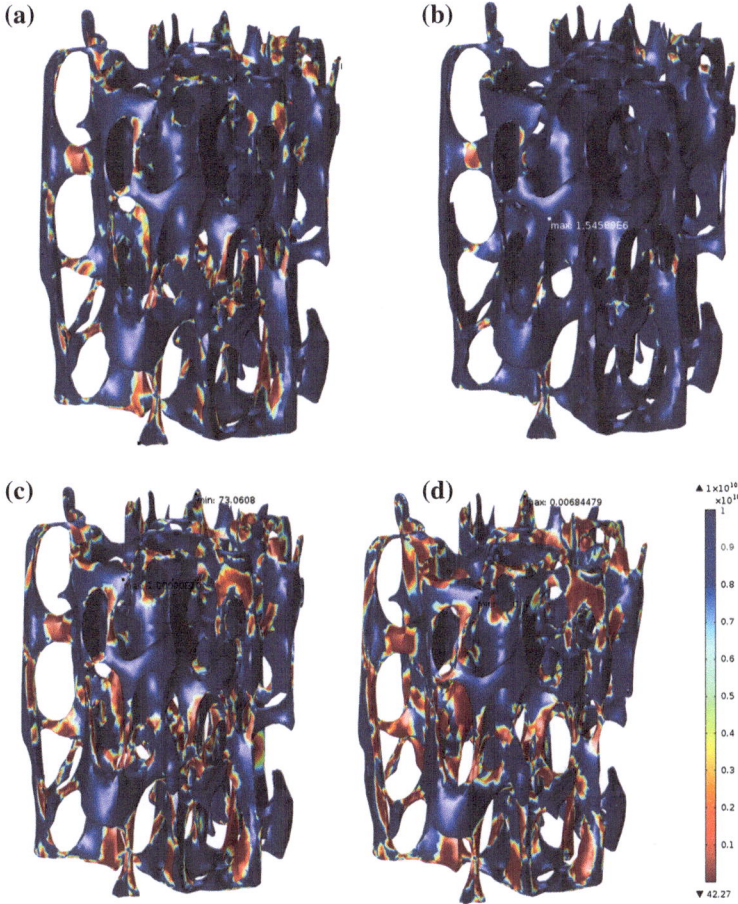

Fig. 4.10 Number of cycles to failure of torsional load. **a** 10 % of total loading. **b** 20 % of total loading. **c** 30 % of total loading. **d** 40 % of total loading

Table 4.8 Number of cycles to failure for torsional load on trabecular bone

Load (%)	10	20	30	40
N_f	611	159	73	42

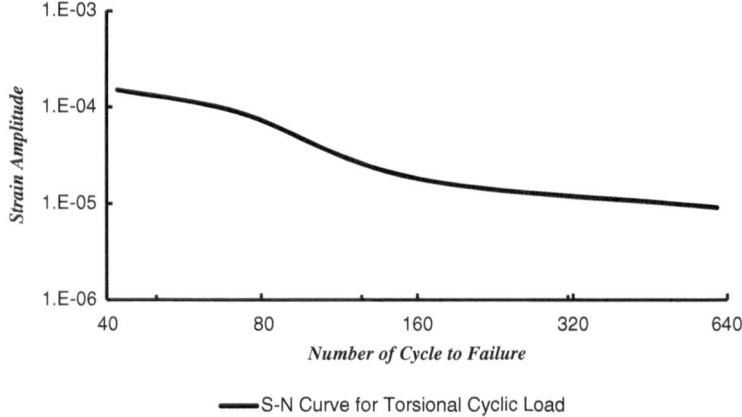

Fig. 4.11 Strain amplitude and number of cycles to failure in torsional load

difference of type of load effects on trabecular bone. In 10 % of load, this value reaches to 611 cycles, and with increase to 40 % of total load it increase to 42 cycles to failure. In addition, the sample selected for this study show sample of trabecular bone which involved with osteoporosis disease and loos its volume fraction (BV/TV) and surface density (BS/TS) or other morphological indices (Table 4.8).

Figure 4.11 show strain amplitude versus number of cycles to failure for torsional load in trabecular bone.

From Fig. 4.11 is clear that with decrease of strain amplitude number of cycles to failure increase and trabecular bone is exactly follow Coffin-Manson law which found in many researches and this fact is verified with simulation of this phenomenon.

4.3.3 Fatigue S-N Curve of Trabecular Bone Subjected to Multi-axial Load

Since trabecular bone had different behaviour in its fatigue life respect to different types of load (torsional and axial), the real situation of load impose on such a structure is combination of axial and torsional load (multi-axial load) which can show the exact behaviour of trabecular bone respect to our physiological activities

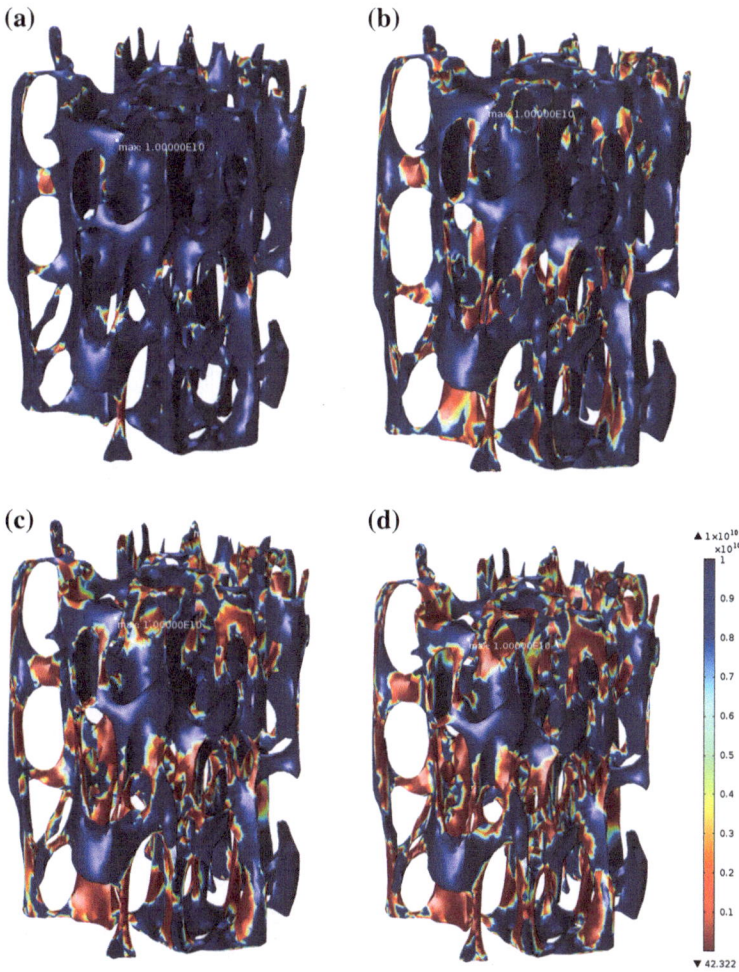

Fig. 4.12 Number of cycles to failure for multi-axial load. **a** 10 % of total loading. **b** 20 % of total loading. **c** 30 % of total loading. **d** 40 % of total loading

and this make it more sensitive than previous models. Figure 4.12 show the multi-axial load imposed on trabecular bone and damage initiated.

In Fig. 4.12, trabecular bone subjected to multi-axial load is presented, the results of damage and failed struts show similarity of such a type of load with torsional load, however, amplitude of damage can be seen more than torsional load due to axial load added in this model. In the 10 % of total load it is clear that just 5 % arch of rod-like trabecular bone failed and damage initiated, however in torsional load it was just arch and rod like but not same amplitude as shown in Fig. 4.13.

The result of fatigue life of trabecular bone subjected to the multi-axial load is tabulated in Table 4.9 and illustrated in Fig. 4.14 as shown below.

The result shows that number of cycle to failure for multi axial load is close to the data and trend of torsional load. In contrast because multi-axial is combination of axial and torsional, consequently all data decrease based on torsional load data for example if in torsional load with 10 % of total load fatigue life of bone was 611 cycles now in multi-axial load with the same load amplitude but fatigue life of trabecular bone 550 cycles.

(a)   **(b)**

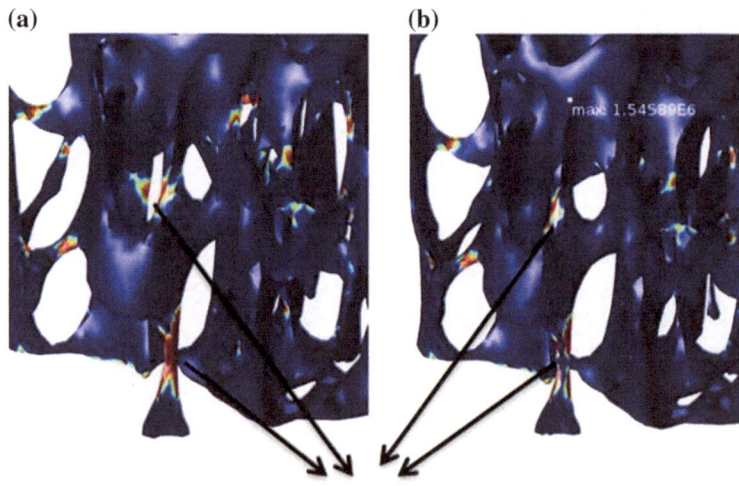

Torsional and multi-axial load effect on damage initiation on struts and growth fast is a function of load

Fig. 4.13 Trabecular bone with failed struts subjected to **a** multi-axial **b** torsional load

Table 4.9 Number of cycles to failure for trabecular bone subjected to multi axial load

Load (%)	10	20	30	40
N_f	550	139	65	38

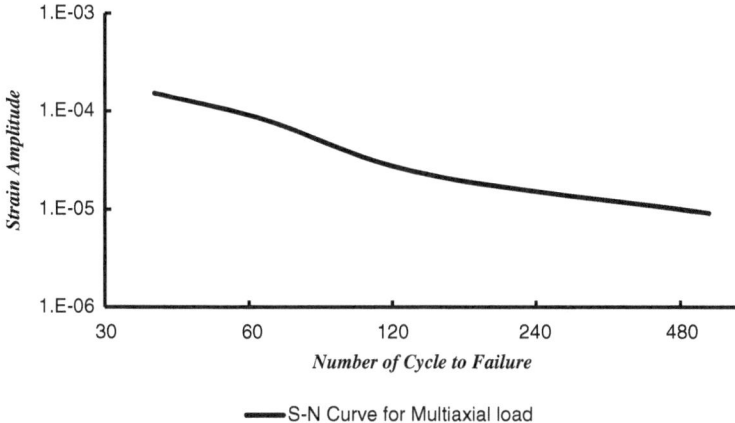

Fig. 4.14 Strain and number of cycle to failure for trabecular bone subjected to multi-axial load

4.4 Summary of Results

In this chapter two analyses had performed, static and dynamic analysis. In static analysis trabecular bone was subjected to axial load two parameter monitored which were stress versus volumetric strain and effective plastic strain that is the crucial factor of damage initiation within structure. In the axial load results were presented plastic strain initiated at 30 % of total load, however, in torsional load and multi-axial load failed structure occur at the first step of cycles and found that struts are not withstand against torsional load. Then, dynamic analysis had been performed for fatigue life estimation. In this step, we applied elastic strain as well as plastic strain as total strain and number of cycle to failure to plot the S-N curve. Results show that trabecular bone subjected to axial load has considerable fatigue life which in 40 % of total load it reaches to 1226.80 cycles approximately, however, this value for torsional load reach to 42 cycles and for multi-axial 38 cycles. Torsional load known as the most effective load, which causes damage within struts and trabeculae. According to physiological activities the most body movement impose load to the hip which combined with axial and torsional.

Chapter 5
Conclusion and Recommendation of Multi-axial Fatigue of Trabecular Bone in Normal Walking

Abstract Analysis of fatigue life of bone support us in various aims such as design artificial trabecular bone fatigue life estimation was performed by applying strain-based method, results show that fatigue is altered based on different type of load; femoral head of bovine trabecular bone was used in this analysis. In dynamic analysis also S-N curve extracted, results show that trabecular bone follow Coffin-Manson law and counted as reliable method for estimation of fatigue life of trabecular bone. Trabecular bone sample subjected to axial, torsional and multi-axial load, has especial stress distribution history within the structure. This finding also clear that, our simulated model has same results in compare with the experimental part. This analysis concerned about load type and its amplitude, however, bone morphological indices is counted as one another important part that can effect on fatigue life of bone. Analysis of different types of load such as axial, torsional and multi-axial in different anatomical sites and analyse its in-phase and out of phase that effect on fatigue life considerably will be expected for those are interested in such area.

Osteoporosis disease makes bone fragility and cause bone have shorter life and close to the fracture in low number of cycles due to physiological activities. Analysis of fatigue life of bone support us in various aims such as design artificial trabecular bone, investigate effect of trabecular bone on implant and select drug based treatment aimed at slowing or stopping bone loss. Fatigue life of trabecular bone is depend on various parameters; different types of load imposed on it, morphological indices such as BV/TV, BS/TS, number of trabeculae in different anatomical sites, load and its direction that makes it close to fatigue life or damage in inner site of bone.

In this analysis fatigue life estimation was performed by applying strain-based method, results show that fatigue is altered based on different type of load; femoral head of bovine trabecular bone was used in this analysis. A sample extracted from this part and subjected to three different loads. Axial, torsional and multi-axial load were imposed on trabecular bone. Effective plastic strain which is a crucial factor of fatigue and damage initiation was computed numerically and results show that

© The Author(s) 2016
M. Mostakhdemin et al., *Multi-axial Fatigue of Trabecular Bone with Respect to Normal Walking*, Forensic and Medical Bioinformatics,
DOI 10.1007/978-981-287-621-8_5

torsional and multi-axial load are same respect to plastic localization which occur in the arch part of trabecular bone. The value of data are close to each other respect to torsional and multi-axial load and consequently, when trabecular faced with torsional load then plastic localization and damage initiation would be occurred quickly by means of which fatigue life of bone will be decrease drastically.

In dynamic analysis also S-N curve extracted, results show that trabecular bone follow Coffin-Manson law and counted as reliable method for estimation of fatigue life of trabecular bone. In such analysis, life of trabecular bone will decrease by increasing load amplitude and plastic strain initiated and growth in arch part of trabeculae. Axial load has no more effect on trabecular as torsional or multi-axial has. In axial load fatigue life reaches to 1226 cycles at 40 % of total load of body, however this value in torsional and multi-axial load decrease to 42 and 38 cycles respectively. Selected bone was simulated as a sample of bone which is involved with osteoporosis disease and more fragile to track the effect of load and stress amplitude on weakens trabeculae. Results also point out that the most sever plastic strain occur in the trabeculae which is perpendicular respect to the load. Since post-processing part of finite element analysis is vast, many factors might be considered to track the fatigue behaviour of bone and analysis of effect of load on rotated trabeculae is beyond the scope of this book. Furthermore, some consideration about this angle is explained to make it more clearly for those who are interested to continue researching on such area.

Trabecular bone sample subjected to axial, torsional and multi-axial load, has especial stress distribution history within the structure. In axial load, sample that include many trabeculae with various rotated angle and faced with such a load had no considerable sever stress or strain amplitude, however, this phenomenon in torsional load is completely different. In torsional load, trabeculaes that are perpendicular to the torsional load had higher stress and strain amplitude than struts which are parallel to the angle of load. These findings show that trabeculae are too weak against transverse load and behave such as brittle material; however, in longitudinal case stress is less than transverse load. This finding also clear that, our simulated model has same results in compare with the experimental part. Previous researchers also found that trabecular bone is fragile against torsional load which was mentioned in literature review.

In axial load, initiated plastic strain was less torsional and multi-axial load respect to load amplitude increment. Multi- axial and torsional load had same behaviour respect to plastic localization and fatigue life curve approximately. This finding shows that torsional load has more effect on failure of inner site architecture than axial load. The point is, if torsional load imposed on sample rotate until become as a transverse load on structure failure occur more quickly than longitudinal case. In addition, struts might be failed after few cycles of load based on obtained results.

This analysis concerned about load type and its amplitude, however, bone morphological indices is counted as one another important part that can effect on fatigue life of bone. When bone faces with osteoporosis disease, volume fraction and bone density will be loosed. Then after, struts will not be able to tolerate against load since they have become weak. Furthermore, this part also can be considered in two types of analysis, first considering bone is dead structure and modelling and re-modelling cannot be done, second modelling and remodelling being done but with low rate, and in this considering a factor of modelling and remodelling which effect on bone life is highly recommended for further study.

As results obtained in this research and some literature that have carried out by other researchers, fatigue life estimation of trabecular bone is accurate in real bone that faced with osteoporosis disease and some cells in bone such as osteoblast and osteoclast has no ability to model and re-model bone and this bone is counted as dead structure. For live bone and estimation of fatigue life, considering model and remodel of bone after every load cycles is highly recommended for further study. Also analysis of different types of load such as axial, torsional and multi-axial in different anatomical sites and analyse its in-phase and out of phase that effect on fatigue life considerably will be expected for those are interested in such area.